DESIGN OF LOW-VOLTAGE LOW-POWER CMOS DELTA-SIGMA A/D CONVERTERS

THE KLUWER INTERNATIONAL SERIES IN ENGINEERING AND COMPUTER SCIENCE

ANALOG CIRCUITS AND SIGNAL PROCESSING

Consulting Editor: **Mohammed Ismail**. *Ohio State University*

Related Titles:

DESIGN OF MODULATORS FOR OVERSAMPLED CONVERTERS, *Feng Wang, Ramesh Harjani,* ISBN: 0-7923-8063-0

SYMBOLIC ANALYSIS IN ANALOG INTEGRATED CIRCUIT DESIGN, *Henrik Floberg,* ISBN: 0-7923-9969-2

SWITCHED-CURRENT DESIGN AND IMPLEMENTATION OF OVERSAMPLING A/D CONVERTERS, *Nianxiong Tan,* ISBN: 0-7923-9963-3

CMOS WIRELESS TRANSCEIVER DESIGN, *Jan Crols, Michiel Steyaert,* ISBN: 0-7923-9960-9

DESIGN OF LOW-VOLTAGE, LOW-POWER OPERATIONAL AMPLIFIER CELLS, *Ron Hogervorst, Johan H. Huijsing,* ISBN: 0-7923-9781-9

VLSI-COMPATIBLE IMPLEMENTATIONS FOR ARTIFICIAL NEURAL NETWORKS, *Sied Mehdi Fakhraie, Kenneth Carless Smith,* ISBN: 0-7923-9825-4

CHARACTERIZATION METHODS FOR SUBMICRON MOSFETs, edited by *Hisham Haddara,* ISBN: 0-7923-9695-2

LOW-VOLTAGE LOW-POWER ANALOG INTEGRATED CIRCUITS, edited by *Wouter Serdijn,* ISBN: 0-7923-9608-1

INTEGRATED VIDEO-FREQUENCY CONTINUOUS-TIME FILTERS: ***High-Performance Realizations in BiCMOS,*** *Scott D. Willingham, Ken Martin,* ISBN: 0-7923-9595-6

FEED-FORWARD NEURAL NETWORKS: ***Vector Decomposition Analysis, Modelling and Analog Implementation,*** *Anne-Johan Annema,* ISBN: 0-7923-9567-0

FREQUENCY COMPENSATION TECHNIQUES LOW-POWER OPERATIONAL AMPLIFIERS, *Ruud Easchauzier, Johan Huijsing,* ISBN: 0-7923-9565-4

ANALOG SIGNAL GENERATION FOR BIST OF MIXED-SIGNAL INTEGRATED CIRCUITS, *Gordon W. Roberts, Albert K. Lu,* ISBN: 0-7923-9564-6

INTEGRATED FIBER-OPTIC RECEIVERS, *Aaron Buchwald, Kenneth W. Martin,* ISBN: 0-7923-9549-2

MODELING WITH AN ANALOG HARDWARE DESCRIPTION LANGUAGE, *H. Alan Mantooth,Mike Fiegenbaum,* ISBN: 0-7923-9516-6

LOW-VOLTAGE CMOS OPERATIONAL AMPLIFIERS: ***Theory, Design and Implementation,*** *Satoshi Sakurai, Mohammed Ismail,* ISBN: 0-7923-9507-7

ANALYSIS AND SYNTHESIS OF MOS TRANSLINEAR CIRCUITS, *Remco J. Wiegerink,* ISBN: 0-7923-9390-2

COMPUTER-AIDED DESIGN OF ANALOG CIRCUITS AND SYSTEMS, *L. Richard Carley,* Ronald S. Gyurcsik, ISBN: 0-7923-9351-1

HIGH-PERFORMANCE CMOS CONTINUOUS-TIME FILTERS, *José Silva-Martínez, Michiel Steyaert, Willy Sansen,* ISBN: 0-7923-9339-2

SYMBOLIC ANALYSIS OF ANALOG CIRCUITS: ***Techniques and Applications,*** *Lawrence P. Huelsman, Georges G. E. Gielen,* ISBN: 0-7923-9324-4

DESIGN OF LOW-VOLTAGE BIPOLAR OPERATIONAL AMPLIFIERS, *M. Jeroen Fonderie, Johan H. Huijsing,* ISBN: 0-7923-9317-1

STATISTICAL MODELING FOR COMPUTER-AIDED DESIGN OF MOS VLSI CIRCUITS, *Christopher Michael, Mohammed Ismail,* ISBN: 0-7923-9299-X

SELECTIVE LINEAR-PHASE SWITCHED-CAPACITOR AND DIGITAL FILTERS, *Hussein Baher,* ISBN: 0-7923-9298-1

ANALOG CMOS FILTERS FOR VERY HIGH FREQUENCIES, *Bram Nauta,* ISBN: 0-7923-9272-8

ANALOG VLSI NEURAL NETWORKS, *Yoshiyasu Takefuji,* ISBN: 0-7923-9273-6

DESIGN OF LOW-VOLTAGE LOW-POWER CMOS DELTA-SIGMA A/D CONVERTERS

by

Vincenzo Peluso
Katholieke Universiteit Leuven

Michiel Steyaert
Katholieke Universiteit Leuven

and

Willy Sansen
Katholieke Universiteit Leuven

KLUWER ACADEMIC PUBLISHERS
BOSTON / DORDRECHT / LONDON

A C.I.P. Catalogue record for this book is available from the Library of Congress.

ISBN 0-7923-8417-2

Published by Kluwer Academic Publishers,
P.O. Box 17, 3300 AA Dordrecht, The Netherlands.

Sold and distributed in North, Central and South America
by Kluwer Academic Publishers,
101 Philip Drive, Norwell, MA 02061, U.S.A.

In all other countries, sold and distributed
by Kluwer Academic Publishers,
P.O. Box 322, 3300 AH Dordrecht, The Netherlands.

Printed on acid-free paper

This printing is a digital duplication of the original edition.

Second Printing 2001.

Contents

List of Symbols and Abbreviations

Physical Symbols

γ	Bulkfactor
μ	mobility
ε_{Si}	Permittivity of silicon
ε_{ox}	Permittivity of silicon SiO_2
ϕ_{GB}	Difference in workfunction between gate and bulk of MOST
ϕ_F	Bulk Fermi potential
C_{ox}	Gate oxide capacitance
k	Boltzmann's constant
q	Elementary charge
Q_{ox}	Fixed oxide charge density in MOS gate
T	Absolute temperature
N_B	Bulk dopant concentration
N_G	Gate dopant concentration
n_i	Intrinsic dopant concentration of silicon
U_t	Thermal voltage $\frac{kT}{q}$

Electrical Symbols

α	Common mode sampling division factor, integrator gain error
β	Integrator pole error
C_i	Capacitance specified as i
C_{eff}	Effective capacitive load
ε_{ox}	Permittivity of oxide
$\overline{di_X^2}$	Average noise current of element X
$\overline{dv_X^2}$	Average noise voltage of element X
$\overline{dV_{out}^2}$	Average output noise voltage
E	Energy [J]
η	Efficiency
f	Frequency [Hz]

f_u	Unity gain frequency
f_T	Transistor unity current gain frequency
$\mathcal{F}_{dc}$	DC feed-back factor
ϕ_i	i^{th} phase in a two phase non-overlapping clock scheme
ϕ_{id}	delayed clock phase i
g_{mi}	Transconductance of
Γ	Excess noise factor
g_{meff}	Effective g_m
g_o	Transistor small-signal output conductance
I	Current
KP	μC_{ox}
KF	Spice $1/f$ noise factor
L_i	Length of MOST i
m	number of stages in cascade $\Delta\Sigma$ modulator
M_i	MOST no. i or no. *ia* and *ib* etc. if present
n	$\Delta\Sigma$ modulator order
q_i	charge stored on capacitor i
P_X	Power of quantity X
R_i	Resistor no. i
R_s	Internal Resistance
R_o	Oversampling ratio
S_i	Switch no. i
V	Voltage
V_D	Voltage drop over diode
V_{BS}	Bulk to Source voltage
V_{GS}	Gate to Source voltage
V_{DS}	Drain to Source voltage
V_{ov}	MOST overdrive voltage, i.e. $v_{GS} - V_T$
V_{DSsat}	Drain to Source saturation voltage of MOST
V_{GST}	MOST overdrive voltage, i.e. $v_{GS} - V_T$
V_R	Ripple Voltage
V_{REF}	Reference voltage in the Original Switched Opamp Technique
$V_{REF,lo}$	Low reference voltage in the Modified Switched Opamp Technique, e.g.,
$V_{REF,hi}$	High reference voltage in the Modified Switched Opamp Technique, e.g.,
$V_{REF,\Delta\Sigma}$	$\Delta\Sigma$ modulator reference voltage
V_T	MOST threshold voltage
W_i	Width of MOST i
ω_u	Unity frequency pulsation
ω_{GBW}	loop GBW product

Abbreviations

ADC	Analog-to-Digital Converter
AGC	Automatic Gain Control
BiCMOS	Process combining Bipolar and CMOS devices

BW	Bandwidth
CMFB	Common Mode Feedback
$CMRR$	Common Mode Rejection Ratio
CMOS	Complementary Metal Oxide Semiconductor
Δ_l	Linear $\Delta\Sigma$ performance degration
Δ_{DR}	Total $\Delta\Sigma$ performance degration
DAC	Digital-to-Analog converter
DR	Dynamic Range
DR_r	Dynamic Range of reference topology
DR_{sl}	Dynamic Range of single loop $\Delta\Sigma$ modulator topologies
DR_{cs}	Dynamic Range of cascade $\Delta\Sigma$ modulator topologies
FFT	Fast Fourier Transform
GBW	Gain-Bandwidth Product
IC	Integrated Circuit
LPF	Low-Pass filter
*n*MOST	n-channel MOSFET transistor
*p*MOST	p-channel MOSFET transistor
OL	Overload Level of $\Delta\Sigma$ modulator
OPAMP	Operational Amplifier
OTA	Operational Transconductance Amplifier
PSD	Power Spectral Density
$PSNR$	Peak Signal to Noise Ratio
$PSRR$	Power Supply Rejection Ratio
SNR	Signal to Noise Ratio

1 INTRODUCTION

1.1 Situation of the Research

In this era where the market for portable electronic systems such as wireless communication devices, hearing aids, consumer electronics, etc. is continuously expanding, there is a growing need for the development of low voltage and low power circuit techniques and system building blocks. Both low voltage operation and low power operation are of great importance for portable applications. Low voltage operation is demanded because it is desirable to use as few batteries as possible for size and weight considerations. Low power consumption is necessary to ensure a reasonable battery lifetime. An example of a standard system building block is an Analog to Digital Converter (ADC), unavoidable as interface between the analog world and the digital signal processing hardware. A very widespread type of ADC in telecommunications applications or audio and voice applications is the $\Delta\Sigma$ modulator. A $\Delta\Sigma$ modulator is suited for low to medium speed but high resolution conversion. Other types of converters exist. Successive approximation and algorithmic converters can perform medium speed and medium accuracy conversion. Examples of high speed but low to medium accuracy converters are flash, interpolating, folding and pipelined converters. Compared to the last two classes of converters whose performance is ultimately limited by matching properties, the $\Delta\Sigma$ ADC has the advantage of being quite insensitive to circuit imperfections such as device mismatch and non-ideal opamp behavior. When designing for very low voltage and low power operation, the latter advantage becomes even more important. The lower the supply voltage, the more difficult it is to maintain specifications such as high gain, to mention just one.

Probably the most robust way of implementation of a $\Delta\Sigma$ ADC is with Switched Capacitor techniques. Their robustness and inherent linearity combined to efficiency of implementation are the reasons for which Switched Capacitor techniques have become as widespread as they are nowadays. Nonetheless, there exist other implementation techniques as well. Also of discrete time nature are the Switched Current (SI) techniques. These have, however, proven to suffer from non-linearity, caused by transistor parasitics. Circuit techniques to alleviate that problem seem to raise the power consumption a lot [Moe 97], without even reaching high performance. SI can also not be called fit for low voltage operation. Their great disadvantage is that they require to switch gate-source voltage levels, which are roughly in the middle between the power rails and this is the worst situation for a switch. Furthermore the circuit techniques to reduce the linearity problems include cascoding, which cannot be done at very low voltage operation. The second technique is an implementation with continuous time filters. These are inherently more sensitive to clock jitter which limits the performance, but high resolution implementations exist. Both $Gm-C$ [Zwa 96b] and active $R-C$ [Mat 94][Mat 97][Zwa 97] filters have been applied. However, achieving very low voltage operation can only be done with a low V_T process.

It would be a great advantage if the high quality Switched Capacitor properties could be kept for low voltage operation. However, when designing Switched Capacitor circuits for lower and lower voltages, quite quickly we run into a severe problem: the switch-driving problem. Classic Switched Capacitor techniques fail to operate for a supply lower than about $2.5V$. The low supply voltage does not allow enough over-drive to turn on the transistors used as switches anymore. It would further be desirable to be able to avoid the use of expensive multi threshold processes that offer a low threshold device.[Baz 95c] The prevalent solution nowadays is to use voltage multiplication to drive the switches.[Au 97][Rab 96][Gri 96] In this respect it is important to realize that there exists another driving force towards low voltage operation of CMOS circuits in general. It is the continuing trend to down scaling of the transistor dimensions. For that reason the future deep sub-micron MOST devices will not be able to withstand the multiplied voltage. The SIA predicts that the maximum supply voltage will drop to $1.5V$ in 2001 and to $0.9V$ in 2009. [SIA 97] It will thus become imperative to avoid the application of voltage multipliers. The Switched Opamp technique offers a solution to this problem [Ste 93][Cro 94], and underlies the realizations presented later in this book.

1.2 Overview of Existing Low–Voltage Low–Power $\Delta\Sigma$ A/D Converters

An overview of the existing Low–Voltage Low–Power $\Delta\Sigma$ A/D Converters is given in Table 1.1. It shows the type, power supply voltage, the dynamic range (*DR*), the signal bandwidth (*BW*) and the power consumption (*P*). The type specifies whether it is a classic Switched Capacitor implementation (SC), with voltage multiplication (VM) or with low threshold devices (LV). An implementation with the Switched Opamp Technique is indicated by 'SO'. A continuous time implementation is indicated by 'CT'. Only converters with a power supply lower than $2.5V$ are considered. In the table the converters are ordered by increasing power supply. A question mark next to the *DR* indicates that the originally reported value is ambiguous. The last column is a figure of merit expressing how good the Power-Resolution-Bandwidth trade-off is. The higher the number, the more power efficient the implementation is.

Ref.	*Type*	V_{DD} (V)	DR (dB)	BW (kHz)	P (μW)	$\frac{4kTDR^2BW}{P}$ ($\times 10^{-6}$)
[Pel 98b]	SO	0.9	77	16	40	330
[Bas 97c]	SO	1	45(?)	20	240	0.04
[Baz 95c]	SC, LV	1	54	4	100	0.16
[Mat 94]	CT, LV	1	58	192	1560	1.3
[Tan 95]	SI, VM	1.2	58(?)	8	780	0.1
[Mat 97]	CT, LV	1.2	94	20	6500	128
[Pel 97c]	SO	1.5	74	3.4	100	14
[Gri 96]	SC, VM	1.8	94	3.5	2000	74
[Rab 96]	SC, VM	1.8	92	25	5400	121
[Au 97]	SC, VM	1.95	73	8	340	8
[Zwa 96b]	CT	2.2	80	3.4	200	28
[Sau 96]	SC, VM	2.4	75(?)	16	1100	8

SC = Switched Capacitor, VM = Voltage Multiplier, LV = Low V_T, SI = Switched Current, CT = Continuous Time

Table 1.1. *Low Voltage, Low Power ΔΣ ADC overview.*

In this work two realizations are presented. The one in [Pel 97c] is the first ΔΣ modulator designed with the Switched Opamp Technique. The one of [Pel 98b] and [Pel 98a] is besides the lowest supply voltage ΔΣ ADC, the lowest supply voltage Switched Capacitor circuit reported in literature. It shows the best figure of merit of all low voltage ΔΣ ADCs.

1.3 Objectives

The objective in this book is to investigate the feasibility of designing ΔΣ ADCs in standard CMOS processes for very low voltage operation combined with low power operation. The Switched Opamp Technique is chosen for the implementation because it is a low voltage Switched Capacitor technique that does not need voltage multipliers and can be implemented in both present and future-deep-sub-micron standard CMOS processes. It is hence at the same time the goal to further elaborate the Switched Opamp Technique by developing dedicated circuitry and improving the state of the art. This concerns the required voltage supply, the power consumption and the performance measured by dynamic range and speed of operation.

A systematic design methodology is followed in the design process. First the ΔΣ modulator topology parameters are optimized on the system level. Then the main circuit non-idealities are modeled and applied in behavioral simulations of the modulator. These provide the required circuit specification in order to ensure that the real system, with non-ideal components, reaches the intended performance. Since this methodol-

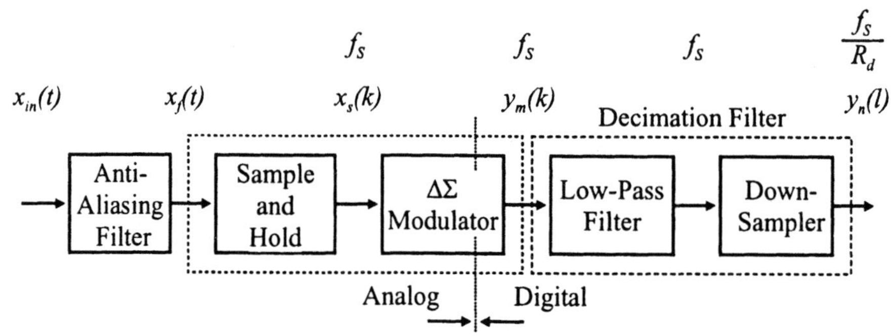

Figure 1.1. *The complete ΔΣ ADC consists of an anti aliasing filter, a sampler, the ΔΣ modulator itself, a digital low-pass filter and a down-sampler. The anti aliasing filter is in most cases very simple. The sampling is inherently performed in a Switched Capacitor filter. The boundary between Analog and Digital is near the output of the ΔΣ. The work in this book is restricted to the ΔΣ modulator itself.*

ogy avoids over estimation of the required circuit specifications, it is an important part of low power design.

Fig. 1.1 shows the complete ΔΣ modulator ADC system. It consists of an anti aliasing filter, a sampler, the ΔΣ modulator itself, a digital low-pass filter and a down-sampler. The anti aliasing filter is in most cases very simple. The sampling is inherently performed in a Switched Capacitor filter. The digital low pass filter and the down sampler are implemented by the decimation filter. The boundary between Analog and Digital is at the output of the ΔΣ modulator for single loop modulators. Cascaded modulators require some more digital circuits to combine the several bit-streams to form the output signal. The work described in this book is restricted to the (analog part of the) ΔΣ modulator itself.

1.4 Outline

The contents of the several chapters is summarized below.

Chapter 2 deals with the ΔΣ topology on the system level. A new implementation method for the classic single loop and cascade ΔΣ modulators with half delay integrators (instead of full delay integrators) is presented. Further ΔΣ topologies are analyzed in the light of a linear model which allows to quantify the performance degradations of higher order single loop modulators and cascade modulators. Ample simulations allow to find the optimal parameters.

Chapter 3 gives an overview of existing low voltage Switched Capacitor techniques, such as the use of low V_T devices and voltage multipliers. A full overview of the present status of the Switched Opamp Technique is given in a historical and at the same time logical order.

Chapter 4 investigates the limitations imposed by the restriction of low supply voltage. Further some existing and novel low voltage building blocks are discussed.

Among other things a new low voltage class AB OTA and common mode feedback amplifier (CMFB) are presented and extensively analyzed.

Chapter 5 discusses design considerations. The first part presents various models taking the main circuit non-ideal characteristics into account. These allow behavioral simulations to be carried out and to fix the required opamp specifications that ensure that the intended performance is met in an implementation. The second part concerns the power consumption. The influence of all the parameters, and especially the low supply voltage, on the power consumption is investigated

Chapter 6 presents two implementations of low voltage, low power $\Delta\Sigma$ ADCs. The first one is a single ended implementation of a $\Delta\Sigma$ ADC operating with $1.5V$ supply voltage and consuming 100 μW for a 74 dB dynamic range in a 3.4 kHz bandwidth. The second implementation operates with 900 mV. It achieves a 77 dB dynamic range in a 16 kHz bandwidth and consumes 40 μW. It is a differential implementation.

Chapter 7 presents the final conclusions.

2
ΔΣ MODULATOR TOPOLOGIES

2.1 Introduction

Analog-to-Digital conversion is an essential function in modern digital signal processing systems. A number of different implementation techniques for A/D converters exist. From these, $\Delta\Sigma$ modulators provide an efficient way of trading off speed for resolution and at the same time are quite insensitive to component mismatches. Therefore $\Delta\Sigma$ modulators have been designed and in use for a long time, and this practice will continue. Depending on the application, different specifications are requested. Some applications require high speed and high resolution converters, but at relaxed supply voltage specifications, while others require very low voltage operation, normally combined with low power operation. Regardless of the destination of the converter, its design should be optimized. The design of any analog system consists of three major steps. The first is the choice of the optimal architecture. The second is the derivation of the specifications of the building blocks and the last is the design of the building blocks on the circuit level such that they achieve the given specs and hence ensure the overall system performance. As far as the architectural aspect of a $\Delta\Sigma$ ADC is concerned, there are several degrees of freedom. There is the choice between the several existing topologies and then there is the choice of the topology parameters. These are the filter coefficients. It is plausible that for a specific topology those parameters that maximize its performance should be chosen. Only when the maximum achievable performance obtainable with a specific topology is known, a funded choice between alternatives can be made.

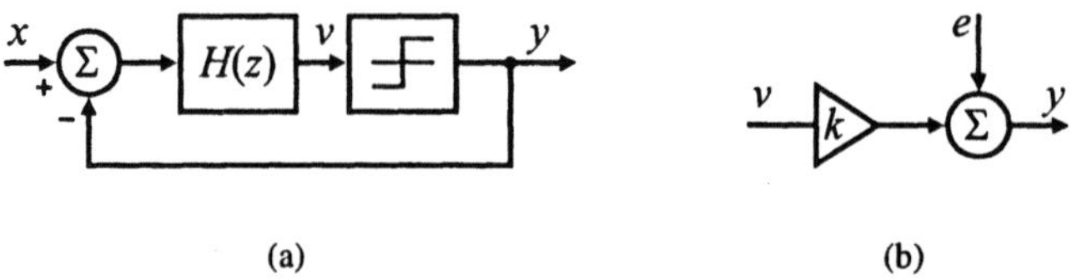

Figure 2.1. *(a) Generalized single bit ΔΣ modulator topology. A single bit ΔΣ modulator consists of a filter and a quantizer in a unity feedback loop. (b) The quantizer model consists of a gain factor plus a noise source.*

In this book the focus is on implementation of ΔΣ modulators using the Switched Opamp Technique. The application of this technique calls for a modification of the ΔΣ modulator using half delay integrators.

Consistent with the issues of the above paragraphs, in this chapter the new ΔΣ modulator topology modification is introduced. Then a topology study is carried out to find the optimal parameters for the classic ΔΣ modulators, i.e., single loop and cascaded. First, a reference ΔΣ modulator is defined in the next section. It serves as a reference to compare the performance of the real modulators with.

2.2 Definitions

2.2.1 A Reference ΔΣ Modulator

A generalized single bit ΔΣ modulator topology is shown in Fig. 2.1. It consists of a loop filter and a 1-bit quantizer in a feedback loop. The conventional analysis method for ΔΣ modulators consists in modeling the quantizer by a gain block with gain k and and a white noise source, that represents the quantization noise process as shown in Fig. 2.1. [Can 92]. Although this model can not describe perfectly the operation of the loop (e.g. stability) nor model some phenomena like the decrease of signal to noise ratio for large input power levels, it gives good results for performance calculation purposes. The output of the generalizedelsigmod is

$$Y(z) = H_x(z)X(z) + H_e(z)E(z) \tag{2.1}$$

where $X(z)$ and $E(z)$ are the input and quantization error signals in the z-domain. The functions $H_x(z)$ and $H_e(z)$ are the signal and the noise transfer functions (*STF*, *NTF*) and are given by

$$H_x(z) = \frac{H(z)k}{1+H(z)k} \tag{2.2}$$

$$H_e(z) = \frac{1}{1+H(z)k} \tag{2.3}$$

The filter gain $H(z)$ is made very large in the signal band. Hence the input signal passes to the output, while the quantization error is strongly reduced. Consider for

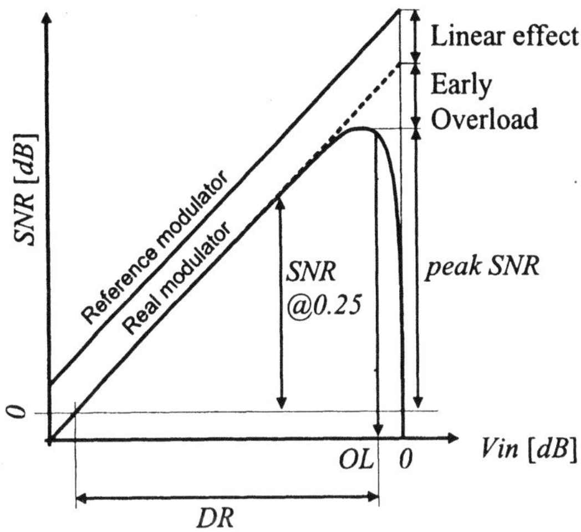

Figure 2.2. *Typical ΔΣ modulator performance graph. With respect to the reference modulator, the performance of a real modulator suffers from a linear degradation due to scaling of the coefficients and a degradation due to early overload for large input signals.*

H(z) an nth order filter with all poles at DC

$$H(z) = \left(\frac{z^{-1}}{1 - z^{-1}}\right)^n \tag{2.4}$$

If the quantizer gain factor is assumed to be $k = 1$ the following results

$$\begin{cases} |H_x(z)| &= 1 \\ H_e(z) &= (1 - z^{-1})^n \end{cases} \tag{2.5}$$

Now the elements are present to define the reference ΔΣ modulator. The purpose of it is to provide a basis for comparison of the performance of the discussed topologies.

> "The hypothetical ideal ΔΣ modulator is defined to have a *STF* and *NTF* defined by (2.5)".

This ΔΣ modulator is an idealized system. It does not suffer from performance degradation due to scaling of the coefficients for stability purposes. Therefore it is also hypothetical, because it cannot exist. Speaking of an *ideal* ΔΣ modulator on the other hand, is not realistic. The gain coefficient in (2.4) has arbitrarily been chosen unity. Choosing it larger than unity would also increase the noise suppression.

2.2.2 *Performance Criteria*

Fig. 2.2 shows the typical ΔΣ modulator performance graph. It is the *SNR* versus the relative input amplitude. Usually both are expressed in dB. The input amplitude is taken relative to the reference voltage. With respect to the ideal, the performance of real ΔΣ modulators suffers from two degradations of different nature. The first is a linear degradation. The topology of Fig. 2.1(a) combined with the filter of (2.4) cannot function as such. It is not stable. As is shown further on, stabilizing the modulator requires reducing the filter gain, which causes the performance to drop linearly with the gain reduction. The second performance reduction is a non-linear effect. It is due to the early overloading of the modulator for large input signals. This makes that the peak *SNR* (*PSNR*) does not occur for the full scale signal, but somewhat lower. The difference between the *SNR* that would result if the overloading effect would not occur and the obtained *PSNR* is the second performance degradation. In order to be able to properly compare the performance of different topologies the following criteria are defined:

$SNR_{-12\,dB}$ The for a relative input amplitude -12 dB (0.25 on a linear scale). A small input signal does not overload the modulator and hence allows to assess the linear degradation separately.

PSNR The gives the maximum performance achievable with a certain topology. It is a measure for the maximum resolution of the converted output signal. It can also be called the output dynamic range.

OL The, is the relative input amplitude for which the structure overloads. For some topologies the overload is quite gradual, therefore it is defined arbitrarily as $max\{(V_{in}) : SNR = PSNR - 6\,dB\}$. This means that a ΔΣ modulator is considered to operate properly for input amplitudes up to the one where the *SNR* decay due to early overload is 6 dB.

A fourth performance quantifier of a ΔΣ modulator is the dynamic range (*DR*). It is here defined as the input dynamic range, i.e., the ratio of the maximum applicable signal power and the quantization noise level. It is not used in the performance comparison in this chapter because it is linearly dependent on the above defined criteria. It can be calculated as

$$DR = SNR_{-12\,dB} + 12\,dB + 20log(OL) \tag{2.6}$$

2.2.3 Performance of the Reference ΔΣ Modulator

The performance criteria defined in the previous section can now be derived analytically for the reference topology, making use of the linearized model of the generalized ΔΣ modulator. The quantization noise source, $e(t)$, is assumed to be a random process with a uniform amplitude distribution in time between $-\Delta/2$ and $+\Delta/2$, and with a white noise spectrum. [Can 92] Δ is the quantizer step size. The quantization error noise power is then

$$e_q^2 = \frac{1}{3}\left(\frac{\Delta}{2}\right)^2 \tag{2.7}$$

The in-band noise power is

$$N_q = \frac{2}{f_s}\int_0^{f_o} \frac{\Delta^2}{12}\left|1-z^{-1}\right|^{2n}_{z=e^{2\pi\frac{f}{f_s}}} df \quad (2.8)$$

$$\approx \left(\frac{\Delta}{2}\right)^2 \frac{1}{3\pi(2n+1)}\left(\frac{\pi}{R_o}\right)^{2n+1} \quad (2.9)$$

where f_o is the signal bandwidth, f_s the sampling frequency, and $R_o = \frac{f_s}{2f_o}$ the over-sampling ratio. If a sinusoidal signal with an amplitude A_x relative to $\Delta/2$ is considered as input, the signal power S and SNR are found as

$$S = \frac{1}{2}A_x^2\left(\frac{\Delta}{2}\right)^2 \quad (2.10)$$

$$SNR_r = \frac{3\pi}{2}A_x^2(2n+1)\left(\frac{R_o}{\pi}\right)^{2n+1} \quad (2.11)$$

If the general case of a B-bit quantizer is considered, the maximum normalized output signal amplitude and the dynamic range of the reference (underscore r) modulator are given by:

$$A_{x,max} = 2^B - 1 \quad (2.12)$$

$$DR_r = \frac{3\pi}{2}\left(2^B-1\right)^2(2n+1)\left(\frac{R_o}{\pi}\right)^{2n+1} \quad (2.13)$$

The latter reduces for a 1-bit quantizer to

$$DR_{r,1b} = \frac{3\pi}{2}(2n+1)\left(\frac{R_o}{\pi}\right)^{2n+1} \quad (2.14)$$

The linear model is intrinsically not capable of modeling the overloading. Hence, the overload level equals unity for all orders:

$$OL_r = 1 \quad (2.15)$$

Table 2.1 lists the *SNR* values corresponding with the hypothetical reference topology. It shows the calculated *SNR* and *PSNR* for oversampling ratios of 16, 32, 64, and 128, for a 1-bit quantizer and a 4-bit quantizer.

The performance of a real modulator differs from the reference by a certain amount. In order to express this performance difference two factors are defined as

Δ_l The performance difference due to linear effects.

Δ_{DR} The combined performance difference due to linear and non-linear (early overload) effects.

The performance of the *real* modulator can then be expressed as

$$SNR = \Delta_l^2\frac{3\pi}{2}A_x^2(2n+1)\left(\frac{R_o}{\pi}\right)^{2n+1} \quad (2.16)$$

$$DR = \Delta_{DR}^2\frac{3\pi}{2}\left(2^B-1\right)^2(2n+1)\left(\frac{R_o}{\pi}\right)^{2n+1} \quad (2.17)$$

n	2			3			4		
R_o	SNR	DR	OL	SNR	DR	OL	SNR	DR	OL
16	37 / 61	49 / 73	1	53 / 76	65 / 88	1	68 / 91	80 / 103	1
32	52 / 76	64 / 88	1	74 / 97	86 / 109	1	95 / 118	107 / 131	1
64	67 / 91	79 / 103	1	95 / 118	107 / 130	1	122 / 146	134 / 158	1
128	82 / 106	94 / 118	1	116 / 139	128 / 151	1	149 / 173	161 / 185	1

Table 2.1. *Calculated performance of the ideal ΔΣ modulator (1-bit / 4-bit quantizer).*

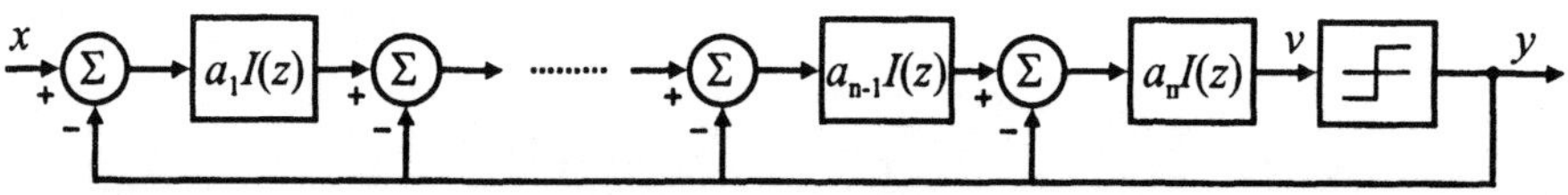

Figure 2.3. *classic n^{th} order single loop ΔΣ modulator topology, with parameters reduced to the minimum number of independent parameters.*

Expression (2.16) is only valid for an input amplitude sufficiently below overload. These performance difference factors are a function of the ΔΣ modulator topology parameters, being the filter coefficients. In order to find an absolute performance difference of a specific topology, the coefficients maximizing Δ_l and Δ_{DR} have to be found.

2.3 Single Loop ΔΣ Modulators

In this section single loop topologies of 2nd to 4th order are discussed. First, however, a new concept of using half delay integrators in ΔΣ modulators is introduced.

2.3.1 *Single Loop ΔΣ Modulator using Half Delay Integrators*

The block diagram of an nth order classic single bit ΔΣ modulator is shown in Fig. 2.3. It is a series of integrators with feedback to each internal node. This is one way of turning the generalized topology of Fig. 2.1 into a stabilizable system. The topology has been reorganized as to have the minimum number of independent parameters. Often the single loop topologies are shown with feedback coefficients different from unity. Application of building block manipulations allows to transform both representations into each other. The feedback coefficients come only into play in a pre-implementation phase, in order to fix the signal ranges on the internal nodes. The internal signal swings are irrelevant in the simulation of theoretical ΔΣ modulators. Also irrelevant for simulation purposes, in fact, is the last integrator coefficient.

The classic single loop ΔΣ modulator of Fig. 2.3 uses full delay integrators, with a transfer function of

$$I(z) = a\frac{z^{-1}}{(1 - z^{-1})} \tag{2.18}$$

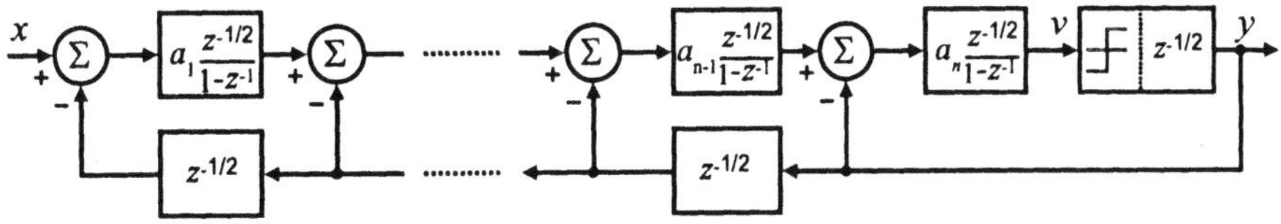

Figure 2.4. *n^{th} order single loop ΔΣ modulator topology using half delay integrators, with parameters reduced to the minimum number of independent parameters.*

The goal in this context is to design ΔΣ modulators applying the Switched Opamp Technique. In Section 3.4.1.2 it is shown that the basic Switched Opamp integrator cell offers only and exactly a half phase delay to the signal, i.e.,

$$I_{\frac{1}{2}}(z) = a\frac{z^{-\frac{1}{2}}}{(1 - z^{-1})} \tag{2.19}$$

If a full delay integrator is required, the integrator needs to be complemented with a delaying buffer. This is an analog function which requires an extra opamp, increasing the silicon area and the power consumption. It would be advantageous to be able to design ΔΣ modulators making use only of the half delay integrators. This requires the availability of ΔΣ modulator topologies built up of half delay integrators. The ΔΣ modulator with full delay integrators can be modified to the topology using half delay integrators of Fig. 2.4 [Pel 98b]. This result is obtained by splitting the integrator function in two parts (a half delay integrator and a half delay block) and by applying standard building block manipulations to the half delay block. The quantizer always takes up a half phase. Consequently, there is no fundamental difference between a realization of a first order topology ($n = 1$) with full and half delay integrators. In the process of shifting the half delay blocks, $n - 1$ half delay blocks appear between the input and the feedback point. These can simply be left out. The result is that the half delay blocks have been shifted to the feedback path. There they are in the digital domain and can be implemented with a latch. Note that the amount of delays around each loop remains unaltered. Only the number of delays between the input and the output has decreased.

Historically a full delay integrator has been called a discrete digital integrator or *DDI* and a half delay integrator a lossless discrete integrator or *LDI*. [Bru 75] In this book the choice is made to use the former names, as they are a more explicit reminder of the property that is of importance in this context: the *full phase delay* versus the *half phase delay*.

2.3.2 Single Loop ΔΣ Modulator Performance

The *STF* and *NTF* of the single loop topology (with half delay integrators) are

$$|H_x(x)| = \left| \frac{\prod_{i=1}^{n} a_i k \left(\frac{z^{-\frac{1}{2}}}{1+z^{-1}}\right)^n}{1 + \sum_{i=1}^{n} \prod_{j=i}^{n} a_j k \left(\frac{z^{-1}}{1+z^{-1}}\right)^{n-i+1}} \right| \approx 1 \quad (2.20)$$

and

$$H_e(z) = \frac{1}{1 + \sum_{i=1}^{n} \prod_{j=i}^{n} a_j k \left(\frac{z^{-1}}{1+z^{-1}}\right)^{n-i+1}} \quad (2.21)$$

$$|H_e(z)| \approx \frac{|1 - z^{-1}|^n}{\prod_{i=1}^{n} a_i k} \quad (2.22)$$

The *NTF* is indifferent of the fact whether a half delay or a full delay integrator is used. The expressions of the *STF* only differ in the number of delays in the numerator, but the magnitude of both approximate unity. Hence there is in reality no difference whatsoever between a modulator using half delay or full delay integrators. The approximation of (2.22) is valid for low frequencies. It is the low frequency behavior which determines the in-band noise suppression.

Redoing the *SNR* calculation of Section 2.2.3 one finds the linear performance degradation coefficient in (2.16) to be

$$\Delta_l = \prod_{i=1}^{n} a_i k \quad (2.23)$$

This factor cannot be calculated unless the quantizer gain is known. It is seen that in the analytical model, as opposed to a simulation, the last integrator coefficient and the quantizer gain are important indeed.

The choice of the quantizer gain remains an open question. In the simplest approach a constant unity gain is arbitrarily attributed to the quantizer. This approach is an over-simplification. The predicted results do not match simulation results. There is indeed no funded reason to assume that the true gain factor k equals unity. Another approach uses one or more linear gains calculated to minimize the mean-square error between the true quantizer and the linear description of the same circuit [Ard 87]. This method gives better results but is quite involved and impractical for real implementation.

In this analysis a pragmatic approach is used, that is inspired on an observation of a peculiar aspect of the operation of a ΔΣ modulator. Since the last integrator coefficient is irrelevant since it is followed by a comparator, changing it does not affect the performance. In order to obtain the same performance with the analytical expression, the quantizer gain, which only exists in the linear model, must vary accordingly. It seems as if the quantizer gain counterbalances the last integrator coefficient. Therefore, instead of defining an arbitrary and independent quantizer gain, in this model the following is done:

> The product of the last integrator coefficient, a_n, and the quantizer gain, k, is assumed to be a constant for whatever value of a_n, and it is here called

effective last integrator coefficient: [Pel 97a][Mar 97][Mar 98b],

$$a_{n,eff} \equiv a_n\, k = C^t \tag{2.24}$$

Thus, the analytical model that is used here makes abstraction of the exact quantizer gain, but considers the product of the last integrator coefficient by the quantizer gain to be constant. It provides a good match between the calculated and simulated power spectral densities, which is relevant for performance comparison purposes. With this approach and the proper choice of parameters (introduced below) the analytical description matches the simulation results with an accuracy of just a few dB for all topologies and for all oversampling ratios.

Expression (2.23) now becomes

$$\Delta_{l,sl} = \prod_{i=1}^{n-1} a_i a_{n,eff} \tag{2.25}$$

Subscript *sl* stands for 'single loop'. In order to choose the value of the effective last integrator coefficient $a_{n,eff}$ a pragmatic approach is used: the cumulative noise power spectral density obtained from transient behavioral simulations is fitted to the one calculated with the analytical linear representation of the system in the frequency domain that is expressed by (2.21). A good fitting for all frequencies (and specially in the low frequency region), and for all single loop topologies, is obtained when

$$a_{n,eff} = 2 \tag{2.26}$$

For input signals much below overload the *SNR* for a single loop (subscript *sl*) modulator can be represented as

$$SNR_{sl} = \Delta_{l,sl}^2 SNR = \left(\prod_{i=1}^{n-1} a_i a_{n,eff} \right)^2 SNR \tag{2.27}$$

Expression (2.27) shows that for high performance, the integrator coefficients must be large. However, increasing the loop parameters worsens the stability of the loop [Eyn 93a]: The OL decreases and the performance degradation due to premature overload increases, and eventually for too large coefficients the loop is simply unstable. Therefore, the increase in performance due to larger coefficients is counteracted by a lower *OL* that results from it. It can be concluded that decreasing or increasing unreasonably the integrator coefficients yields less than optimal performance. The objective is to find the combinations of coefficients that ensure maximal dynamic range. Note that the linear model of (2.27) shows that equi-performance surfaces are hyperboloidae. This behavior can be verified in practice, when the conditions of the linear model are satisfied, i.e., for input signal levels far below overload.

Extensive *SNR* and *PSNR* simulations have been performed for second to fourth order loops. In Fig. 2.5 the results of *PSNR* simulations for third and fourth order loops, respectively, as function of the first and second integrator coefficients are presented for $R_o = 32$. The 4th order topology is obtained with the third integrator gain equal to

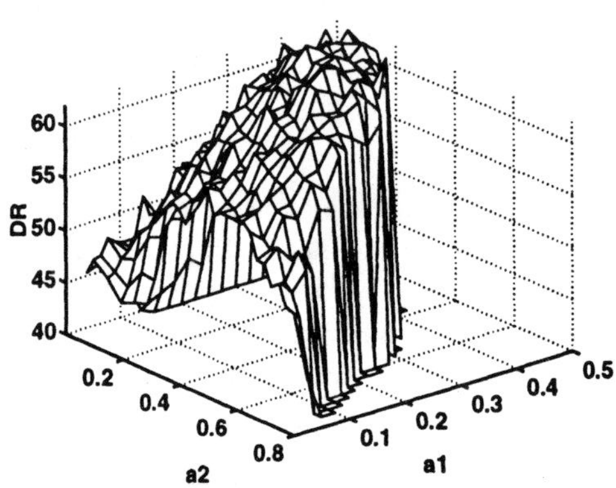

(a) 3rd order surface plot

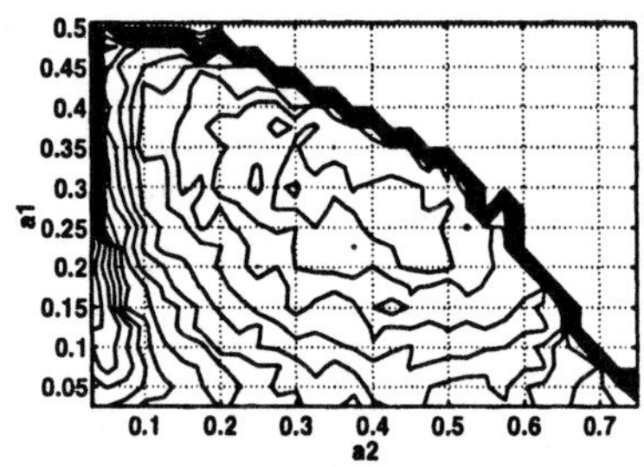

(b) 3rd order contour plot

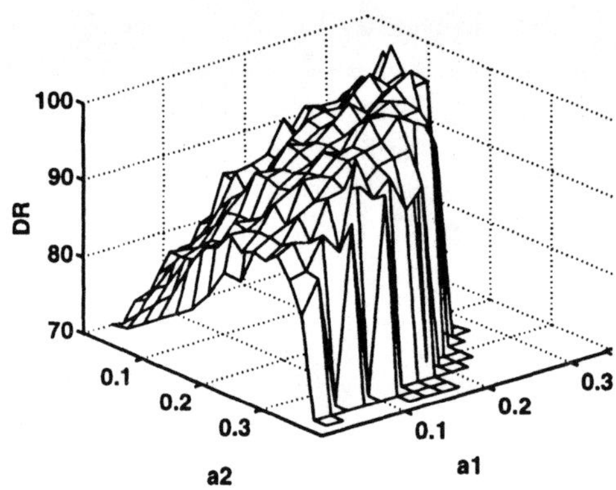

(c) 4th order surface plot

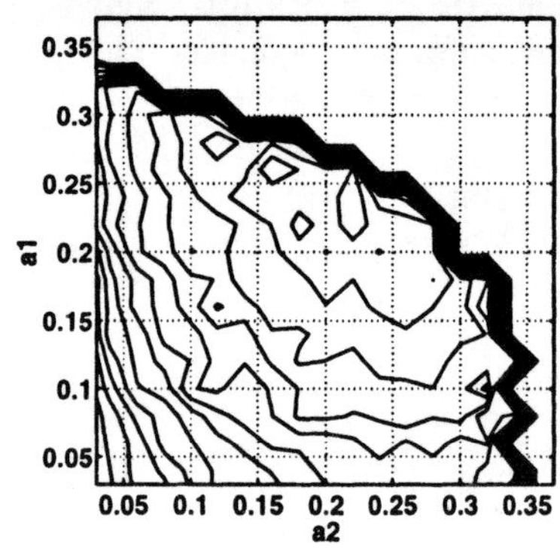

(d) 4th order contour plot

Figure 2.5. *Surface and contour plots of PSNR of the 3rd and 4th order single loop ΔΣ modulators. For parameter combinations much below the stability boundary, the performance increases linearly. There the equi-performance lines are hyperbolae. Near the stability boundary a flat optimum occurs.*

0.5, which is the best value found. Fig. 2.5(a) and (c) are three dimensional plots of the dynamic ranges, illustrating the previously described behavior. Fig. 2.5(b) and (d) are contour plots of the dynamic ranges, on which the stability boundary can clearly be seen, as well as the regions with equal dynamic range. Within the region of stable operation, the points of equal performance lie on hyperbolae as predicted by the linear model (2.27). The flat region of maximal performance, close to the stability border, illustrates that the coefficients should be maximized, but not made so large such that the performance drops too much due to *OL* shift to a lower level. Summarizing, within the boundaries of stability, a constant product of integrator gains yields approximately equal performance, while the maximum performance is to be found close to the stability border.

Based on these simulations and considerations, it can be concluded that the dynamic range of the loops is maximized when the performance difference factor is approximately:

$$\Delta_{l,sl} = \quad 1, \frac{1}{5}, \frac{1}{25} \quad \text{for} \quad n = 2, 3, 4 \tag{2.28}$$

In order to make an intelligent choice for loop parameters some more issues need to be addressed. It is desirable to have the loop coefficients increasing from the first to the last integrator. That way the internal error accumulations are smallest and therefore the internal range of the state variables is decreased. Moreover, in the second order loop the first coefficient must be smaller than 3/4 for stability reasons, implying the same restriction to the one but last coefficient of these topologies. In [Eyn 93a] it was already known that for the fourth order loop, 1/2 is the best choice for stability. Based on these considerations the coefficients shown in Table 2.2 are found to be the optimal ones for single loop topologies. Also shown in the same table are the overload level, dynamic range, signal to noise ratio, and between parenthesis the degradation in performance compared with the ideal topology.

Par	$(a_1, a_2) = (0.5, 0.5)$			$(a_1, a_2, a_3) = (0.2, 0.5, 0.5)$			$(a_1, a_2, a_3, a_4) = (0.2, 0.2, 0.5, 0.5)$		
n	2			3			4		
R	*SNR*	*DR*	*OL*	*SNR*	*DR*	*OL*	*SNR*	*DR*	*OL*
16	35 (-2)	42 (-7)	0.70	38 (-15)	43 (-22)	0.60	34 (-34)	40 (-40)	0.60
32	50 (-2)	57 (-7)	0.70	60 (-14)	65 (-21)	0.55	67 (-28)	71 (-36)	0.60
64	66 (-1)	74 (-5)	0.70	80 (-15)	86 (-21)	0.55	91 (-31)	100 (-34)	0.55
128	80 (-2)	88 (-6)	0.65	105 (-11)	108 (-20)	0.55	123 (-26)	128 (-33)	0.55

Table 2.2. *Topology parameters and performance found by simulation for second to fourth order single loop ΔΣ modulators. Between brackets the difference with the ideal model is shown.*

2.3.3 *Multibit Topologies*

While in the last section the single bit single loop ΔΣ modulator topologies were studied, in this section the multi bit topologies are investigated. As can be concluded based on (2.13), the principal motivation to use multi bit topologies is the fact that this structure offers a direct improvement over one bit topologies of $20log(2^B - 1)dB$.

When linear analysis is applied, the noise transfer function can be calculated and approximated and the performance difference factor calculated:

$$\Delta_{l,mb} = \prod_{i=1}^{n} a_i \tag{2.29}$$

The subscript *mb* stands for 'multi bit'. The first thing that catches the eye is that the quantizer gain does not play a role any longer, because it is now better defined. The system is now intrinsically more stable. Consequently, the integrator coefficients can be increased, which results into an additional benefit, compared with the single loop topologies.

Another important asset of the multi bit modulator is the possibility of realizing higher order noise shaping without any extra circuitry to solve the stability problem [Che 95]. The block diagram of a straightforward implementation of a multi bit $\Delta\Sigma$ modulator is equal to the block diagrams of the single bit $\Delta\Sigma$ modulators shown in Fig. 2.3, but with the one bit quantizer replaced by a multi bit quantizer. Such a straightforward implementation, however, imposes tough demands on the linearity of the multi bit DAC in the feedback path, since the feedback signal is directly subtracted from the input node. Consequently the non-linearity errors cannot be distinguished from the input signal and are not reduced by the feedback loop. Hence the DAC must have a linearity as good as the wanted final resolution, albeit that it only has small number of bits. This poses severe difficulties on the implementation of the DAC. Several attempts to relax the problem modifying in one way or another the classical topology, have been made in the past [Bra 91][Les 90][Hai 91].

There also have been some attempts trying to really eliminate the problem, using in fact a straightforward multi bit approach but addressing the problem of the linearity of the DAC. There are several schemes proposed including digital correction [Cat 89] [Nej 93], and dynamic element matching techniques (including random, clocked, individual level, and data weighted averaging [Car 89][Pla 79][Leu 92][Che 95].

An extensive coverage of all these topologies for all orders is out of the scope of this book. However, for certain applications the multi bit topologies may provide interesting solutions. Hence, just the ideal topology is studied, although keeping in mind that the DAC linearity problem has to be solved and that it implies always some performance degradation.

Since the quantizer gain is now more accurately defined, classical linear analysis can now be used more reliably to verify the stability of the loop for a certain parameter set, under the condition that the quantizer is not driven outside its input range (here considered normalized to ± 1). This means that a_n cannot be made too large.

The optimum parameter sets, and the corresponding performance, found through behavioral simulation for second to fourth order modulators with a 4-bit quantizer are presented in Table 2.3

Note that for the multi bit topologies the product of the integrator coefficients is different from the one found for the single bit topologies, and for optimum performance is equal to:

$$\prod_{i=1}^{n} a_i = \frac{27}{16}, \frac{27}{32}, \frac{27}{128} \quad \text{for} \quad n = 2, 3, 4 \tag{2.30}$$

The first of these actually gives a performance better than the hypothetical reference topology. Note that these are the optimum parameters found using a 4-bit quantizer and for a different number of bits, the optimum parameters may change slightly.

	$(a_1,a_2)=(0.75,2.25)$			(a_1,a_2,a_3) $=(0.5,0.75,2.25)$			(a_1,a_2,a_3,a_4) $=(0.25,0.5,0.75,2.25)$		
n	2			3			4		
R	*SNR*	*DR*	*OL*	*SNR*	*DR*	*OL*	*SNR*	*DR*	*OL*
16	66 (+5)	78 (+5)	0.90	75 (-1)	86 (-2)	0.85	78 (-13)	89 (-14)	0.80
32	80 (+4)	94 (+6)	0.90	97 (0)	107 (-2)	0.85	104 (-14)	117 (-14)	0.85
64	96 (+5)	106 (+3)	0.90	117 (-1)	128 (-2)	0.85	134 (-12)	144 (-14)	0.85
128	110 (+4)	123 (+5)	0.90	139 (0)	145 (-6)	0.85	158 (-15)	168 (-17)	0.85

Table 2.3. *Topology parameters and performance found by simulation for the multi-bit structure. Between brackets the difference with the ideal model.*

2.4 Cascaded ΔΣ Modulators

2.4.1 Cascaded Topologies using Half Delay Integrators

As it is proven possible in Section 2.3.1 to make use of integrators with a half delay instead of a full delay, it is possible to build cascade topologies with them.

The cascaded 2-1 topology with half delay integrators is shown in Fig. 2.6. A comparator takes up a half phase. When using normal full delay Switched Capacitor integrators, the output is sampled during the first half phase and then applied to the comparator. The output sample used to feed to the subsequent stage is taken in the second half phase. When half delay opamps in the Switched Opamp Technique are used, the latter is not possible, because the opamp is then not operating. The sample to be fed to the subsequent stage, must now also be taken a half phase earlier and then delayed a half phase, so it can be properly combined with the comparator output signal. Consequently the scaling factor applied to v now incorporates a half phase delay. The digital logic needed to properly combine the two output bit-streams turns out to be not any different from a full delay integrator implementation. Linear analysis of these structures allows to express the output signal as a combination of the input signal and the quantization noise of the last stage

$$Y_{2-1}(z) = z^{-\frac{5}{2}}X(z) + \frac{1}{c_1}\left(1-z^{-1}\right)^3 E_2(z) \tag{2.31}$$

The cascaded 2-2 topology with half delay integrators is shown in Fig. 2.7. Note that the path from y_1 to y has one and a half delays instead of two in the full delay case. This is because the second stage now only takes up one and a half delay. Again, the digital logic is the same as for full delay integrators, and the output is

$$Y_{2-2}(z) = z^{-3}X(z) + \frac{1}{c_1}\left(1-z^{-1}\right)^4 E_2(z) \tag{2.32}$$

Fig. 2.8 shows the cascade 2-1-1 for which the output is

$$Y_{2-1-1}(z) = z^{-\frac{7}{2}}X(z) + \frac{1}{c_1c_2}\left(1-z^{-1}\right)^4 E_3(z) \tag{2.33}$$

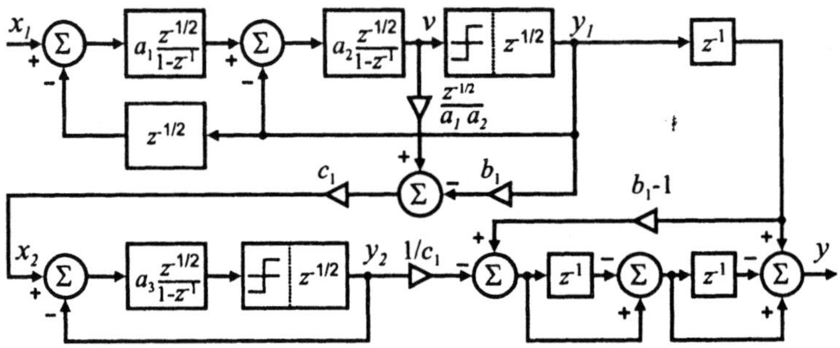

Figure 2.6. *Cascaded 2-1 ΔΣ topology with half delay integrators.*

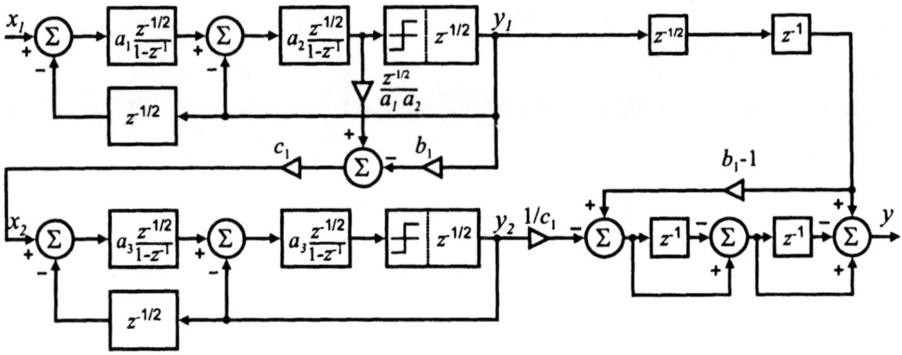

Figure 2.7. *Cascaded 2-2 ΔΣ topology with half delay integrators.*

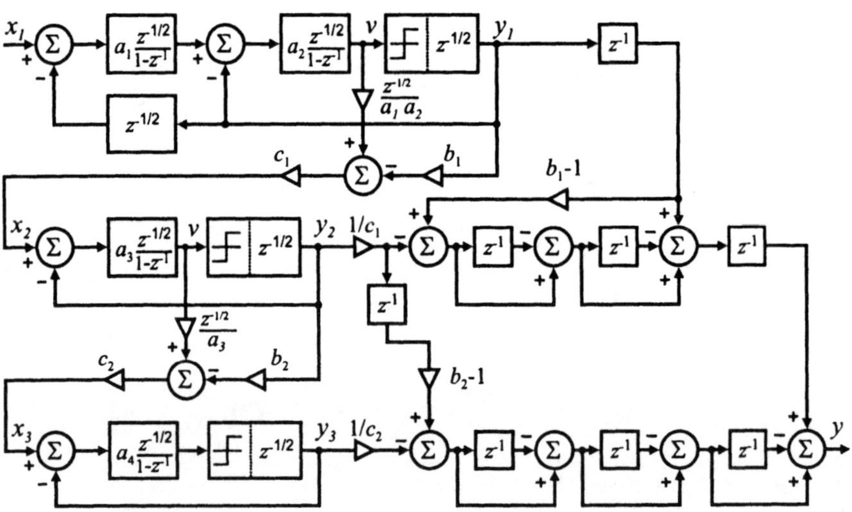

Figure 2.8. *Cascaded 2-1-1 ΔΣ topology with half delay integrators.*

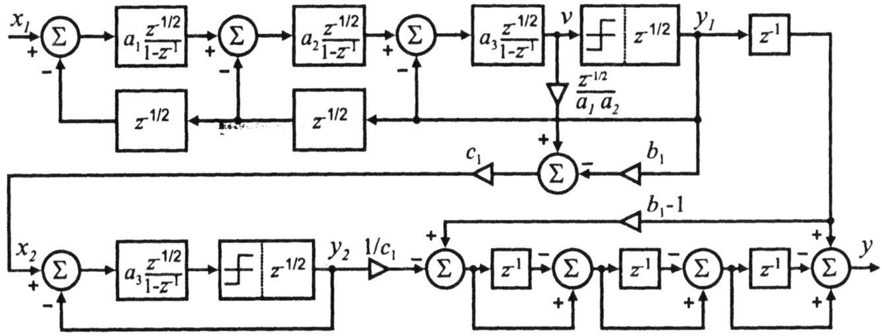

Figure 2.9. *Cascaded 3-1 ΔΣ topology with half delay integrators.*

The cascade 3-1 architecture is also studied [Mar 97][Mar 98b]. The reason why it is analyzed is because it presents an alternative for implementation of a 4th order ΔΣ modulator and it is not possible to state off hand whether it is going to have better or worse performance than the alternatives. It has also not been implemented yet. It is not clear why. Its topology with half delay integrators is shown in Fig. 2.9. The output is expressed as

$$Y_{3-1}(z) = z^{-3}X(z) + \frac{1}{c_1}\left(1 - z^{-1}\right)^4 E_2(z) \tag{2.34}$$

From (2.31)(2.32)(2.33)(2.34) the general form of the noise transfer function is

$$H_e(z) = \frac{\left(1 - z^{-1}\right)^n}{\prod_{j=1}^{m} c_j} \tag{2.35}$$

Taking this *NTF* into account in the *SNR* in the linear analysis of Section 2.2.3 the *SNR* can be represented as

$$SNR_{cs} = \Delta_{l,cs}^2 SNR = \prod_{j=1}^{m} c_j\, SNR \tag{2.36}$$

where *SNR* is that of the reference topology, m is the number of sections and c_j are the coupling coefficients. Subscript *cs* stands for 'cascade modulator'. The performance degradation now depends on the coupling coefficients c_j. The objective is to determine the coefficients a_i, b_j, c_j, in order to maximize the achievable dynamic range.

2.4.2 Choice of the Parameters

The basic idea of the cascaded topologies is to measure the quantization noise, convert it and subtract it from the output. The key problem is to measure the quantization noise. Only the output bit-stream and the state variable at the comparator input can be used, and it is not at all obvious how exactly they should be combined. This is now systematically addressed.

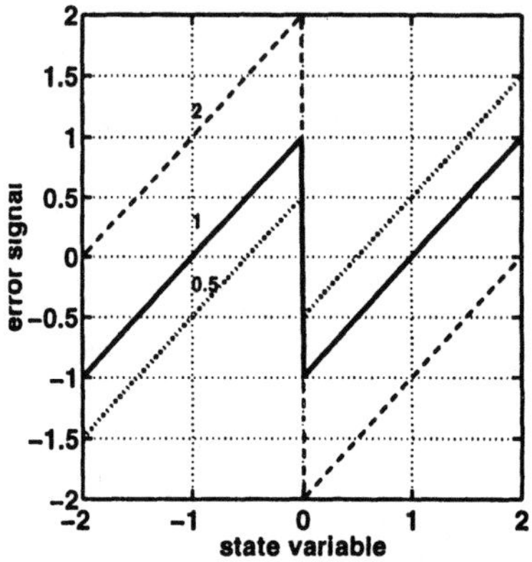

Figure 2.10. *First order ΔΣ modulator error signal. If a scaling is applied before subtraction, the error signal range is not minimal.*

The optimal parameters a_i of each separate modulator of a cascade topology are those found for single loop structures in Section 2.3.2.

The signal in front of the quantizer must always be normalized in such a way that the overall gain from the modulator output (through the feedback loop, filter and scaling) to the subtraction point is unity [Rib 91][Mat 87][Yin 94b]. Hence the scaling coefficient between the quantizer input and the subtraction point should be equal to the inverse of the product of all the real integrator coefficients of that stage: $\prod_{i=1}^{n} a_i$

The factors b_j and c_j are found from the following analysis. The key idea is that the input range of the subsequent stage must be maximally exploited. This means two things. First, the amplitude distribution of the measured quantization noise must be minimized. Second, this distribution noise needs to be matched to the input range of the subsequent stage. This is the role of the scaling coefficients c_j. If the former is not accomplished, the measured error signal must be scaled down more, i.e., a smaller c_j is needed. Expression (2.36) shows that in that case the performance reduces with that amount. The procedure is established by analyzing a first order ΔΣ modulator, with a unity integrator coefficient. For input $x \in [-1, +1]$ and the output $y \in [-1, +1]$, the state variable $v \in [-2, +2]$. Then the direct subtraction of the state variable from the output is between $y - v \in [-1, +1]$. Quite important is the fact that if any scaling is applied before subtraction, the difference has a higher range: $\forall b \in R_0^+ : (by - v) \in [-r, +r], \quad r = max(b, |b - 2| > 1)$ In Fig. 2.10 this behavior is graphically illustrated, where the error signal (state variable minus a scaled version of the output) is shown as function of the state variable itself, for several scaling factors.

In the general case of an nth order modulator the last integrator coefficient a_n is not unity. The signal being subtracted is scaled by the last integrator coefficient and then multiplied by the inverse of the product of all integrator coefficients yielding an overall scaling factor of $a_n \frac{1}{\prod_{i=1}^{n} a_i} = \frac{1}{\prod_{i=1}^{n-1} a_i}$ (see e.g., Fig. 2.6). If the last but one state variable $v_{n-1} \in [-1, 1]$ then the scaled last integrator state variable $v_{n,s} \in [-2\frac{1}{\prod_{i=1}^{n-1} a_i}, +2\frac{1}{\prod_{i=1}^{n-1} a_i}]$. Therefore this overall scaling factor, equal to the inverse of the product of all integrator coefficients but the last, then must be also applied to the output through the b_j factor. Hence b_j should be chosen as

$$b_j = \frac{1}{\prod_{i=1}^{n-1} a_{ji}} \tag{2.37}$$

where j is the stage index and i is the integrator index, counting the first integrator of stage j as 1 and the last integrator of that stage n.

As seen above, with the derived selection of b_j, the error signal is normalized to a range equal to the inverse of the product of the integrator coefficients except the last, in practice (much) smaller than unity. In order to make this range equal to the input range of the subsequent stage a scaling factor must be applied, equal to

$$c_j = \kappa OL \frac{1}{\prod_{i=1}^{n-1} a_{ji}} \tag{2.38}$$

Ideally κ is supposed to be unity. However, in higher than 2nd order stages, due to accumulation effects in the first integrators, the input to the last integrator is sometimes slightly higher than unity, and hence the error signal can also have a slightly higher range. Hence $\kappa < 1$. Since the overload level of the following stage is always smaller than unity, c_j has always to be chosen about equal or smaller than the product of the integrator coefficients except the last. Furthermore c_j should preferably be a power of two in order to simplify the posterior digital scaling. This provides a good starting point to choose c_k, but the best coefficient among a few alternatives strongly depends on the overload level of the next stage, and the optimum value has to be found resorting to simulations.

In order to verify the above considerations experimentally, the distribution of the measured error signal can be calculated from simulation data. It is most clearly demonstrated with the case of the 3-1 cascaded ΔΣ modulator, because that case is less likely to be coincidental. There the distribution of the measure of the error signal using exactly the coefficients obtained as explained above, with $\kappa = 1$ and $OL = 1$, is shown. It is strictly between ± 1, and remarkably flat.

The analysis is not any different for modulators with full or half delay integrators, since the stage couplings and digital signal processing to be applied to the bit-streams are in essence identical. Hence the results of one case are identical to the other case. Following the previously described procedure all the topology parameters for the 2-1, 2-2, 2-1-1, and 3-1 are chosen. Extensive simulations of the loop performance varying all parameters have shown that the former procedure provides indeed the parameters that yield a maximum dynamic range, with the exception of c_j where a few alternatives must be checked by simulations. Nevertheless in some structures it is possible to find other parameter sets that provide comparable performance, but not better. In Table 2.4 and Table 2.5 the *OL*, *PSNR*, *SNR*, and also the degradation in performance (presented between parenthesis) compared with the ideal topology are presented for the 2-1 and 2-2 topology, for two parameter sets. It can be seen that the simulation results for c_1 equal to 0.25 and 0.5 show virtually the same performance. However, for the former

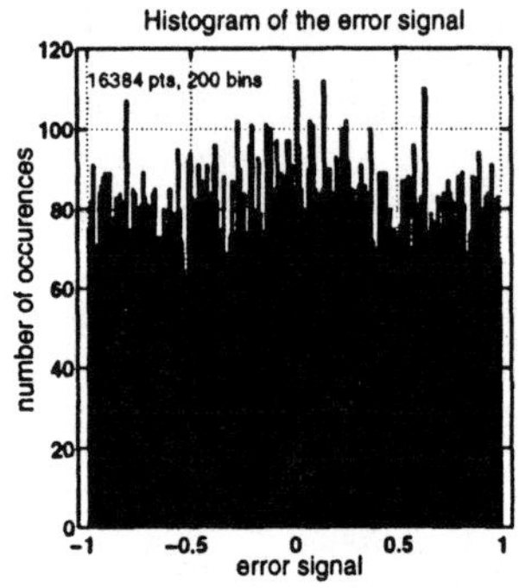

Figure 2.11. *Histogram of the measure of the error signal in 3-1 cascaded ΔΣ modulator. A proper combination of the quantizer output and input through the coefficients b_j and c_j yields a quite flat error signal distribution in the minimum width possible [-1,+1].*

value the overload levels are slightly higher, which is an expected result since the range of the input signal to the second stage is smaller.

Par	$(a_1, a_2, a_3, b_1, c_1) = (0.5, 0.5, 0.5, 2, 0.5)$			$(a_1, a_2, a_3, b_1, c_1) = (0.5, 0.5, 0.5, 2, 0.25$		
R	*SNR*	*DR*	*OL*	*SNR*	*DR*	*OL*
16	46 (-7)	56 (-9)	0.75	43 (-10)	54 (-11)	0.85
32	66 (-8)	77 (-9)	0.70	63 (-11)	76 (-10)	0.85
64	89 (-6)	96 (-11)	0.70	85 (-10)	96 (-11)	0.85
128	112 (-4)	119 (-9)	0.65	108 (-8)	119 (-9)	0.85

Table 2.4. *Topology parameters and performance found by simulation for the 2-1 cascaded structure. Between brackets the difference with the ideal model.*

Par	$(a_1, a_2, a_3, a_4, b_1, c_1) = (0.5, 0.5, 0.5, 0.5, 2, 0.5)$			$(a_1, a_2, a_3, a_4, b_1, c_1) = (0.5, 0.5, 0.5, 0.5, 2, 0.25)$		
R	SNR	DR	OL	SNR	DR	OL
16	58 (-10)	64 (-16)	0.70	55 (-13)	63 (-17)	0.80
32	85 (-10)	92 (-15)	0.70	82 (-13)	90 (-17)	0.80
64	112 (-10)	119 (-15)	0.65	109 (-13)	120 (-14)	0.80
128	141 (-8)	144 (-17)	0.60	138 (-11)	144 (-17)	0.80

Table 2.5. *Topology parameters and performance found by simulation for the 2-2 cascaded structure. Between brackets the difference with the ideal model.*

In Table 2.6 the performance results obtained for the 2-1-1 topology are presented. The parameter set on the left obeys the guidelines. Surprisingly another combination of parameters that offers the same performance exists, namely b_1 equal 3, with the other parameters unchanged.

This is striking because such a good performance is not expected for these parameters. When b_1 equals 3 (instead of 2), too much output signal is being subtracted, and after scaling the input signal to the second stage is normalized to ± 1.5 (instead of ± 1). This means that this stage is being overloaded. One possible explanation might be that the statistics of the input signal to the second stage have a self stabilizing effect.

Par	$(a_1, a_2, a_3, a_4, b_1, c_1, b_2, c_2) = (0.5, 0.5, 0.5, 0.5, 2, 0.5, 1, 1)$			$(a_1, a_2, a_3, a_4, b_1, c_1, b_2, c_2) = (0.5, 0.5, 0.5, 0.5, 3, 0.5, 1, 1)$		
R	SNR	DR	OL	SNR	DR	OL
16	62 (-6)	72 (-8)	0.70	62 (-6)	72 (-8)	0.80
32	90 (-5)	99 (-8)	0.70	88 (-7)	99 (-8)	0.80
64	117 (-5)	125 (-9)	0.65	116 (-6)	126 (-8)	0.80
128	141 (-8)	152 (-9)	0.65	144 (-5)	152 (-9)	0.80

Table 2.6. *Topology parameters and performance found by simulation for the 2-1-1 cascaded structure. Between brackets the difference with the ideal model.*

In Table 2.7 the results obtained for the 3-1 topology are shown. Two parameter sets that yield approximately the same performance are given. Both parameter sets obey the guidelines for the coupling between stages. On the left side the parameter set used for the first stage is the one found in section 2.2 for the single loop third order structure. On the right side the parameter set used is different, but still very close to the condition established by (2.28) - remember that the effective last integrator coefficient equals two. Note that, in this topology, the subsequent digital scaling is not a power of two, but the right side case (factor 1/12) requires only slightly more complicated digital hardware.

Par	$(a_1, a_2, a_3, a_4, b_1, c_1) = (0.2, 0.5, 0.5, 0.5, 10, 0.1)$			$(a_1, a_2, a_3, a_4, b_1, c_1) = (1/4, 1/3, 1/2, 1/2, 12, 1/12)$		
R	SNR	DR	OL	SNR	DR	OL
16	44 (-24)	50 (-30)	0.60	42 (-26)	52 (-28)	0.70
32	71 (-24)	80 (-27)	0.55	71 (-24)	81 (-26)	0.65
64	98 (-24)	107 (-27)	0.55	100 (-22)	109 (-25)	0.65
128	131 (-18)	132 (-29)	0.55	128 (-21)	137 (-24)	0.60

Table 2.7. *Topology parameters and performance found by simulation for the 3-1 cascaded structure. Between brackets the difference with the ideal model.*

2.5 Systematization of the Results

In this section the former results are systematized. However only the signal to noise ratio and the dynamic range are considered.

Expression (2.27) reveals that the single loop performance is degraded by the factors Δ_{sl}, which for optimum performance are given by (2.28). Hence a performance degradation of 0 dB, -14 dB, and -28 dB is expected for the second to fourth order single loop topologies. These results compare well with the *SNR* degradation values presented in Table 2.2, meaning that the differences are less than 6 dB.

Expression (2.29) reveals that the multibit topology performance is altered by a factor equal to the product of all the integrator coefficients, which, for optimum performance is given by (2.30). Compared with the defined ideal topology performance a difference of +5 dB, -1 dB, and -14 dB, is expected for the second to fourth order modulators. The *SNR* degradation values presented in Table 2.3 never differ by more than 2 dB.

Expression (2.36) reveals that the performance is degraded by the digital scaling factor following the last stage. Depending on the topology the scaling factor used is 1, 1/2, 1/4, 1/10, or 1/12, yielding 0 dB, -6 dB, -12 dB, -20 dB, -22 dB. The *SNR* degradation values, presented in tables 4 to 7, never differ more than 4 dB from these values.

These results are summarized in Table 2.8, where a *SNR* degradation coefficient Δ is presented for each topology, both as a linear coefficient and as a reduction in dB. The reduction in dynamic range is difficult to model. Hence the *PSNR* degradation coefficient presented in Table 2.8 is obtained from the previous simulations. This coefficient includes the effect of the *SNR* degradation and also the effect of premature overloading of the structure. Note that in this section only one parameter set for each topology is considered, and between the alternatives the one that for comparable performance has a higher overload level is selected.

n	Type	Parameters	Δ_l	Δ_l [dB]	Δ_{tot} [dB]
2	SL	(a1,a2) = (0.5,0.5)	1	0	-6
3	SL	(a1,a2,a3) = (0.2,0.5,0.5)	1/5	-14	-21
4	SL	(a1,a2,a3,a4) = (0.2,0.2,0.5,0.5)	1/25	-28	-36
3	CS 2-1	(a1,a2,a3,b1,c1)=(0.5,0.5,0.5,2,0.25)	1/4	-12	-12
4	CS 2-2	(a1,a2,a3,a4,b1,c1)=(0.5,0.5,0.5,0.5,2,0.25)	1/4	-12	-16
4	CS 2-1-1	(a1,a2,a3,a4,b1,c1,b2,c2) =(0.5,0.5,0.5,0.5,3,0.5,1,1)	1/2	-6	-8
4	CS 3-1	(a1,a2,a3,a4,b1,c1)=(1/4,1/3,1/2,1/2,12,1/12)	1/12	-22	-26
2	MB 4-b	(a1,a2)=(0.75,2.25)	27/16	+5	+5
3	MB 4-b	(a1,a2,a3)=(0.5,0.75,2.25)	27/32	-1	-2
4	MB 4-b	(a1,a2,a3,a4)=(0.25,0.5,0.75,2.25)	27/128	-14	-15

Table 2.8. *Linear (Δ_l) and total (Δ_{tot}) performance degradation of all topologies compared with the ideal reference topology.*

Fig. 2.12 presents the generalization of the optimum *SNR* and *PSNR* values obtained for all studied topologies as function of the oversampling ratio.

2.6 Comparison of Studied Topologies

The graphs in Fig. 2.12(a) and (b) and especially the last one are useful for anyone who intends to design a $\Delta\Sigma$ modulator with the classic single loop or cascaded topologies. It provides a summary of all those topologies at a glance. Alternatives for a given specification can be chosen and trade-offs can be made. In this section some conclusion are drawn upon the relative performance of the several topologies, based on the graphs depicted in Fig. 2.12.

Comparing the single loop topologies with cascaded topologies generally, the former are always much worse than the latter in terms of *PSNR*. Therefore if the main goal is high resolution, the cascaded topologies offer a much better solution than single loop topologies. This, however, comes at the cost of a much higher sensitivity to non idealities of the building blocks [Rib 91], which may seriously limit their real performance. On the other hand, if a certain performance

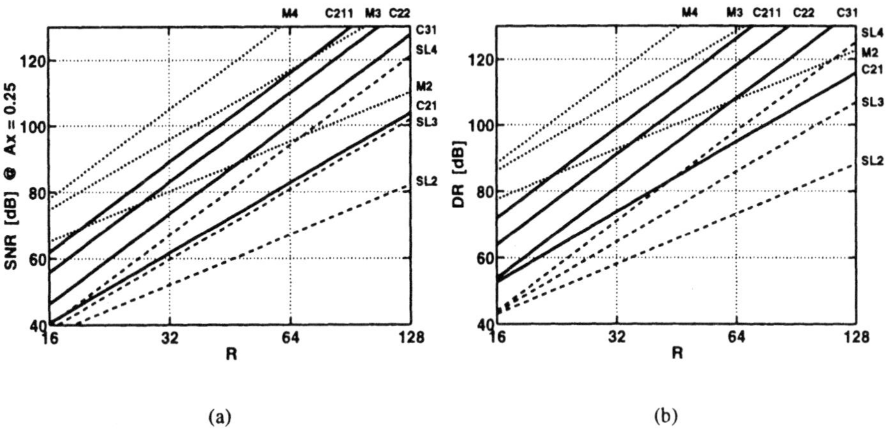

Figure 2.12. *SNR (a) and PSNR (b) v_s oversampling ratio for a normalized input signal amplitude of 0.25 and for all studied topologies. SNR versus oversampling ratio. "SLi" = i^{th} order single loop modulator, "Mi" = i^{th} order multi bit modulator, "Cijk"= i-j-k cascaded modulator.*

is desired it is perhaps simpler with respect to non-idealities to implement for example a fourth order single loop than a cascaded 2-1 (compare the two graphs for R_o higher than ca. 48).

Looking only at fourth order cascaded topologies, it can be seen that the dynamic range of the 2-1-1 structure is better than the 2-2 and the 3-1 by 8 and 18 dB, respectively. Again in this case this higher performance comes at a cost of a higher sensitivity to mismatch in the loop parameters and non idealities of the building blocks [Yin 94b]. However the much lower achievable dynamic range of the 3-1 structure for the same order indicates that this structure most likely does not represent an interesting alternative for practical realizations.

Comparing the cascade topologies with the multi bit topologies, it is obvious from the graph that the most interesting region to explore the multi bit topologies is the region of low oversampling ratios. In this region even a 4-bit second order modulator offers performance comparable to a fourth order single bit cascade modulator. This is logical since the multi bit quantizer provides for all oversampling ratios an intrinsic *SNR*, to which the benefits of oversampling and noise shaping add.

2.7 Conclusions

A modification of the classic ΔΣ modulator topologies which uses half delay integrators is presented [Pel 98b]. Further systematic study of single bit and multi bit single loop and single bit cascaded and ΔΣ modulators is discussed [Pel 97a][Mar 97][Mar 98b]. It allows to find the best coefficient sets for each topology with respect to *PSNR*, as well as the corresponding maximal performances. The performance degradation of single loop ΔΣ modulators relatively to a hypothetical reference topology is explained and quantified: in order for the higher order single loop ΔΣ modulator to be stable, the coefficients must be reduced and the performance degradation is basically equal to the product of the coefficients. Guidelines for the choice of the coupling coefficients in cascaded structures are developed: the input signal to a following stage, which is a measure for the quantization noise, should be calculated in the way that its amplitude distribution has minimal bounds, and it should be scaled to fully exploit but not overload the following stage. Finally, the results are generalized graphically as a function of the oversampling ratio, offering a handy overview of the studied topologies and of trade-offs provided by those.

3 THE SWITCHED OPAMP TECHNIQUE

In this chapter the Switched Opamp Technique is treated. As a starting point the problems of low voltage Switched Capacitor signal processing circuits are considered. The key problem is the driving of the switches. The existing solutions for it are briefly covered: they are voltage multipliers and the use of low V_T processes. Then the Original Switched Opamp principle is introduced. Next the evolution in this field is described. The endpoint is the Differential Modified Switched Opamp integrator cell. Common mode control techniques are presented. A problem in Switched Opamp, is the input stage. This problem and solutions for it are treated last.

3.1 The Problem of Low Voltage Switched Capacitor Signal Processing

Switched Capacitor circuits make use of one basic building block, namely the integrator. It is constructed of three building elements: An opamp or OTA, a switch, and a capacitor. With these, a variety of integrator functions can be made: inverting or non-inverting, delaying or non-delaying, lossy or non lossy. Fig. 3.1 shows a non-inverting, delaying and non lossy integrator cell. The switches S_1 through S_6 are clocked with two phases ϕ_1 and ϕ_2. C_S is the input sampling capacitor, C_{INT} is the integrating capacitor and C_L is the load capacitor and in fact makes out the sampling capacitor of the next integrator.

Lowering the supply voltage of such a circuit has implications on the operation of some of the building elements. The functional property of a capacitor, namely its capacitance, is in fact independent of the supply voltage. opamps and switches, however, are strongly affected. The former need to cope with much less available voltage drop over each transistor. The latter always need a minimal overdrive voltage in order to assure a certain on-resistance. It is possible to design opamps with quite low supply voltage. This matter is treated in more detail in Chapter 4.

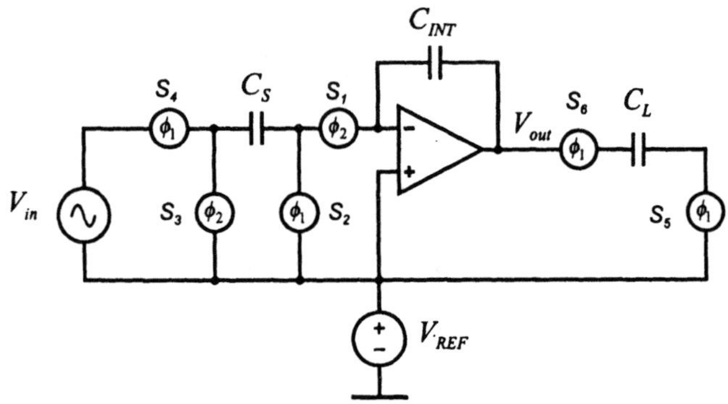

Figure 3.1. *The non inverting Switched Capacitor integrator.*

On the other hand, voltage up-converters are often used to generate a larger supply voltage for the opamp power supply as well, in the case the available supply voltage is really too low.

The most problematic issue, however, in low voltage switched capacitor circuit design is the switch driving problem. In classic Switched Capacitor circuits a complementary switch is preferably used. It is shown in Fig. 3.2. The *n*MOST and *p*MOST on-conductance in this configuration and in settled condition is given by the following expressions [L&S a], where all voltages are referred to V_{SS}.

$$G_{sw,n} = \left[KP_n \left(\frac{W}{L} \right)_n (V_{DD} - V_{in} - V_{Tn}) \right] \tag{3.1}$$

where

$$V_{Tn} = V_{Tn0} + \gamma_n \left(\sqrt{|2\Phi_{F,n} - (-V_{in})|} - \sqrt{|2\Phi_{F,n}|} \right) \tag{3.2}$$

$$G_{sw,p} = \left[KP_p \left(\frac{W}{L} \right)_p (V_{in} - |V_{Tp}|) \right] \tag{3.3}$$

where

$$V_{Tp} = V_{Tp0} - \gamma_p \left(\sqrt{|2\Phi_{F,p} - (V_{DD} - V_{in})|} - \sqrt{|2\Phi_{F,p}|} \right) \tag{3.4}$$

Assuming that $V_{DD} - V_{SS} > V_{Tn} + V_{Tp}$ the *n*MOST conducts for an input signal from V_{SS} on up to V_{Tn} below V_{DD}. The *p*MOST conducts from V_{Tp} on up to VDD. In Fig. 3.3 (a) the simulated switch conductance $G_{sw} = G_{sw,n} + G_{sw,p}$ for a $5V$ power supply and for minimal switch dimensions of $1\ \mu m/0.7\ \mu m$ is shown. The turn on points of the switches are very strongly influenced by the bulk factor γ. In the middle region both transistors conduct in parallel and they can be sized in order to yield a reasonably constant conductance as is shown in Fig. 3.3 b).

The behavior of the switch is of great importance in switched capacitor circuits due to two reasons. Firstly because it may limit the signal range. A complementary switch allows to pass a rail-to-rail input signal. This benefit is lost when lowering the supply voltage as is explained further on. Secondly because it puts a limit on the obtainable frequency of operation. This is because the ON-resistance of the switch together with the sampling capacitor define an RC time constant. In the non-inverting switched capacitor integrator of Fig. 3.1 the input is first sampled

onto the sampling capacitor C_S. Assuming a zero internal resistance of the signal source, the settling of the sampling is completely determined and limited by this RC time constant. In the subsequent integration phase the settling process can only go slower than this due to the finite gain bandwidth product of the OTA.

Although expressions 3.1 and 3.3 indicate that lower switch resistance and thus faster settling can be achieved by upscaling of the device, there is a limit to this caused by the increased capacitive load, the clock feed-through and charge injection. High speed switched capacitor circuits have to cope with these effects and their design is a trade-off between speed and clock feed-through. As the supply voltage is lowered, this problem becomes more severe. Consequently low voltage switched capacitor circuits will in general have reduced speed performance.

The conductance of the complementary switch for the condition $V_{DD} < V_{Tn} + V_{Tp}$ is shown in Fig. 3.4 for $V_{DD} = 1.5V$. In this case there exists an input signal range for which neither of the complementary type switches is turned on anymore. Consequently the input signal cannot be sampled onto the capacitor any longer, and rail-to-rail operation of the switched capacitor circuit is no longer possible. However, a reduced signal range can still be switched if the signal is located in a range where one of the switches is conducting. Because the other switch is then never on, it can be omitted. The best device to choose for application as a single switch is that device for which the KP and V_T ensure the lowest on-resistance. KP_p can be from 2.5 to 4 times lower than KP_n. In most present day CMOS processes the V_{Tp} is either higher than or

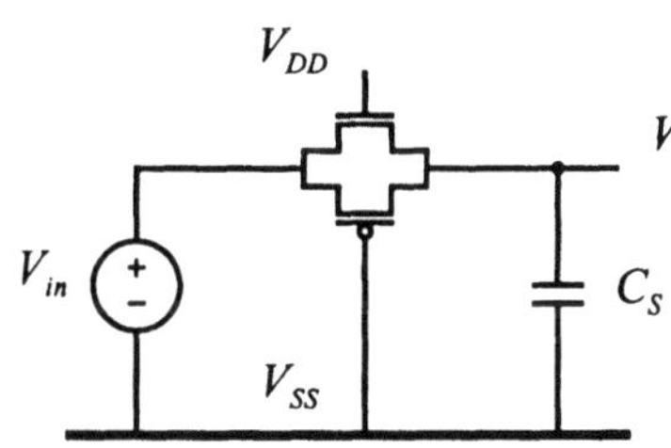

Figure 3.2. *The complementary switch.*

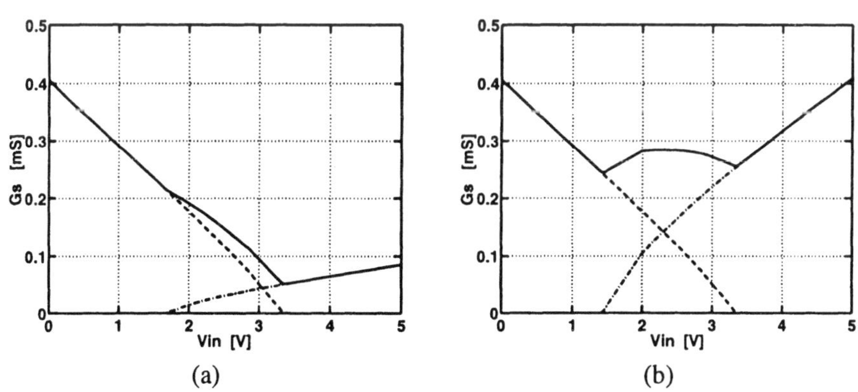

Figure 3.3. *The conductance of* (a) *a minimal size ($W_n/W_p = 1\ \mu m/1\ \mu m$) and* (b) *a properly scaled ($W_n/W_p = 1\ \mu m/4\ \mu m$) complementary switch for a supply voltage of $5V$ in a $0.7\ \mu m$ CMOS process.*

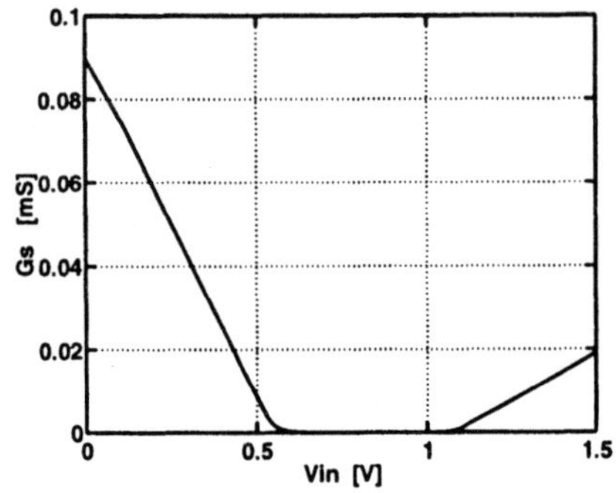

Figure 3.4. *The conductance of a minimal size (W_n/W_p = 1 μm/1 μm) complementary switch for a supply voltage of* 1.5V *in a* 0.7 μm *process.*

comparable to V_{Tn}, but never significantly lower enough to make up for the difference in KP. Therefore it is safe to say that the nMOST makes the best single transistor switch.

Although it is possible to design classic low voltage switched capacitor circuits using a single device switch, as pointed out in the last paragraph, their reduced signal range is a too large restriction. Three techniques have been developed to overcome this problem. These items are treated next. The first one consists of the use of voltage multipliers. The second one is to use MOST devices with reduced threshold voltages. And the third one, which is explored and developed in this book, is the Switched Opamp Technique.

3.2 Voltage Multipliers

As pointed out in the former section, when designing switched capacitor circuits for low supply voltages, ensuring the operation of the opamps and switches becomes problematic. One viable way to effectively cope with the low supply is to apply voltage conversion. This technique uses circuits which convert the voltage from the available supply to another one. The conversion factor can be smaller or larger than unity, positive or negative. In low voltage applications the conversion factor is of course larger than unity. The up-converted voltage is usually a multiple of the available supply voltage and that is why the term "voltage multiplier" is used. In the next sections some examples of a classic and a more recent voltage multipliers are given. Their most important characteristics such as efficiency are treated.

3.2.1 Existing Voltage Multipliers

Various circuits have been proposed and used in literature to implement voltage multipliers. Originally monolithic voltage multipliers stem from non volatile memory circuits where a relatively high potential is needed to write or erase information. This is the case for the Dickson voltage multiplier [Dic 76] which is an improved version of the Cockroft-Walton multiplier [Coc 32] and is much less sensitive to stray capacitors which is important in monolithic implementation. The Dickson multiplier has become a classic and has been often used since [Kru 83] [Wit 89]. A more recent voltage doubler, from DRAM applications is found in [Nak 91]. It looks like it is becoming a classic as well since it has been used in e.g., [Cho 95] and improved in [Fav 97][Fav 98].

Since about a decade voltage multipliers have appeared out of low voltage applications, typically for hearing aids [Cal 89]. In low voltage Switched Capacitor applications Voltage

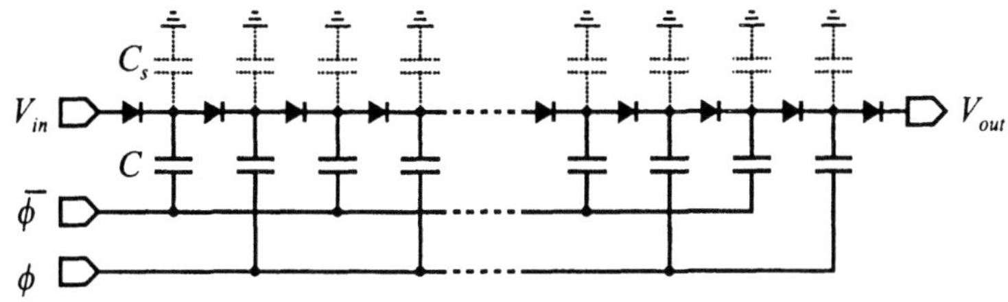

Figure 3.5. *The Dickson Voltage Multiplier.*

Multipliers are being used both to supply a larger voltage for the opamps and for the switches, which often have different voltage multipliers with different final voltage. An example can be found in [Way 92]. The converter supplying the upconverted voltage for the opamps is in fact a simple voltage inverter. The available supply is sampled in one phase with a capacitor that then charges a larger capacitor with a negative voltage the next phase. Thus twice the original voltage becomes available, which is enough to drive the opamps. The switches need yet a higher voltage so another and different voltage converter doubles the obtained voltage again. Other recent examples can be found in [Nic 94] and [Sen 97]

Next, two types of voltage multipliers are treated to get a feeling for the problems and trade-offs that exist.

3.2.2 The Dickson Multiplier

The circuit of Fig. 3.5 is known as the *Dickson Multiplier* [Dic 76]. The operation of the circuit is as follows. A capacitor with a capacitance of C is coupled to each internal node of the diode chain between input and output. The other terminals of the capacitors are alternately coupled to the phases ϕ or $\bar{\phi}$. For a given capacitor, when its clock line goes high, it charges the following one whose clock line is low. In the next half phase, that particular capacitor whose clock line is now low, is being charged by the preceding one. The diodes prevent charge flowing back to the previous capacitor. In the condition that no charge is removed from the system at the output, the system would theoretically evolve to a state where each capacitor is charged to a voltage V_ϕ higher than the preceding one. In practice, however, each capacitor is charged approximately to a voltage $V_\phi - V_D$. The diode current at small forward bias voltage over it, cannot compensate the charge needed to load the stray capacitors C_s. When a load is applied at the output, per phase there is a charge flow from each capacitor to the next, equal to the charge flow into the load. In fact the system acts like some sort of a bucket-brigade line. The difference is that the buckets are not emptied. Each bucket holds more liquid than the previous one, and only a little bit of liquid is transfered to the next bucket, to compensate the consumption at the end of the line.

The output voltage of the circuit is calculated as

$$V_{out} = V_{in} - V_D + N[V'_\phi - V_D] - \Delta_{V_{out}} \tag{3.5}$$

where

V_{in} is the input voltage of the circuit,

V_D is the voltage drop over the diode,

N is the number of stages,

$$V'_\phi = \frac{C}{C + C_s} V_\Phi \quad \text{is the voltage swing of node } i\text{, with} \tag{3.6}$$

V_ϕ the clock swing, and

$$\Delta_{V_{out}} = \frac{N}{f(C + C_s)} I_{out} \quad \text{is the ripple on the output voltage due to the output current.} \tag{3.7}$$

In these expressions, C is the useful capacitance, C_s is the stray capacitance and f is the clock frequency.

Besides the up-converted output voltage, the relevant characteristics of a voltage multiplier are

- current capability
- internal resistance
- output ripple (which is related to the above), and
- the efficiency.

The output ripple V_R is caused by the fact that the load current slightly discharges the last capacitor. It is given by

$$V_R = \frac{I_{out}}{C + C_s} \tag{3.8}$$

This lost charge will be supplied in the next phase by the previous capacitor which will undergo the same voltage drop. N such ripples between input and output cause the voltage drop $\Delta_{V_{out}}$ of (3.7). This equation actually defines the internal resistance of the multiplier:

$$R_S = \frac{N}{f(C + C_s)} \tag{3.9}$$

Efficiency of any kind of transformer is a very important specification, even more so in low power applications. Therefore it is remarkable that no attempt to calculate an expression for the efficiency of the Dickson multiplier can be found in open literature. For this reason and for the purpose of gaining insight in the pro's and con's of applying voltage multipliers in low voltage, low power applications, this exercise is done here. The complete derivation can be found in Appendix A.1.

Efficiency is defined as the ratio of useful power to total dissipated power. The useful power expressed by P_u is the the power delivered to the load and P_l is any power lost. The efficiency is then expressed as

$$\eta = \frac{P_u}{P_u + P_l} = \frac{1}{1 + \frac{P_l}{P_u}} \tag{3.10}$$

The ratio of the lost and useful power can be calculated to be (Appendix A.1)

$$\frac{P_L}{P_U} \approx \left(\frac{N}{V_{out}} \left(\frac{\alpha}{(1+\alpha)} V_\phi^2 \frac{C f}{I_{out}} + (1+\alpha) \frac{I_{out}}{C f} \right) + (1+\alpha) \frac{N V_D}{V_{out}} \right) \tag{3.11}$$

where α is defined as $C_s = \alpha C$.

	C	α	f	V_ϕ	N
Ripple V_R	$\uparrow$	$\simeq$	$\uparrow$	-	$\downarrow$
Source Resistance R_S	$\uparrow$	$\simeq$	$\uparrow$	-	$\downarrow$
Efficiency due to C_s η_{C_s}	$\downarrow$	$\downarrow$	$\downarrow$	$\downarrow\downarrow$	$\downarrow$
Efficiency due to R_S η_{R_S}	$\uparrow$	$\simeq$	$\uparrow$	-	$\downarrow$

Table 3.1. *Influence of the design parameters on the performance of the Dickson Multiplier. The sign $\uparrow$ means that the parameter heading the column should be made large to improve the property stated in the first column on that line. The sign $\downarrow$ means it should be made small, $\simeq$ means no large influence exists and '-' means there is no influence.*

There are three phenomena influencing the efficiency of the voltage multiplier and each cause a term in (3.11).

- The most visible phenomenon is caused by the internal resistance. The physical reason for it is the dissipated energy due to charge redistribution of the charging current flowing from a certain capacitor to the next. It is quantified by the second term.
- The second phenomenon is less visible and easily overlooked: it is the energy lost in the charging and discharging process of the stray capacitors. There is work done during each phase by the clock source to make the current flow, that charges a stray capacitor. This work is lost the next phase, when it discharges again. It is quantified by the first term.
- The third phenomenon is the energy loss that the charges, flowing towards the output to deliver the output current, undergo due to the voltage drop over the diodes.

A good voltage multiplier has low ripple V_R, low source resistance R_S and high efficiency η. Table 3.1 shows the influence of the system parameters on these characteristics for a constant output current I_{out}. From Table 3.1 the following conclusions can be drawn.

- The first column shows that the capacitors need to be large for good performance. Large capacitance means low R_S and thus low efficiency loss due to R_S. The immediate obvious consequence is a *large area consumption* for monolithic voltage multipliers. However, there is an optimum value since increasing the capacitor increases the losses in the stray capacitors which scale with C.
- The second column says that the type of capacitor used should be the one with the least stray capacitance. This has a large influence on the efficiency loss due to the stray capacitance.
- As a consequence of the two above items, it can be concluded that the best performing voltage multipliers will not be monolithic, but will use an *off-chip capacitor*. These capacitors can be made really large and have virtually no stray capacitance. The disadvantage is of course the pin cost.
- The switching frequency f should be high to reduce efficiency loss due to R_S, but it should be low to reduce efficiency loss due to C_s. For a given load current I_{out}, there exists an *optimum frequency* f. So the load current should be known in the design phase. Another solution would be to incorporate a control system with the multiplier that changes f as a function of I_{out}. For off chip capacitors f will be very high.
- The clock amplitude should be small to reduce efficiency loss due to C_s. In order to achieve a certain output voltage the number of stages N should be increased. From the last column

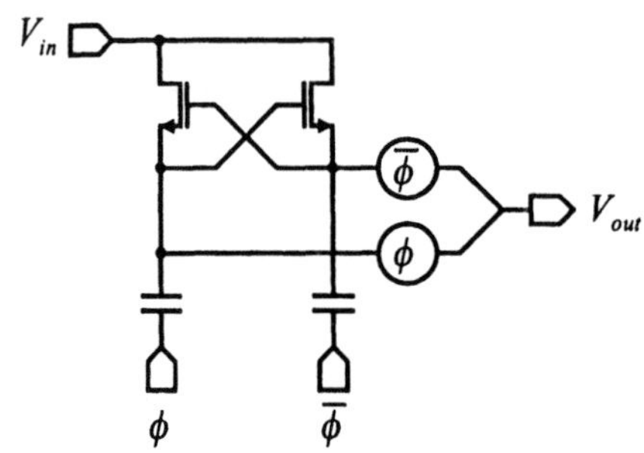

Figure 3.6. *A Voltage Doubler.*

is it clear that N should be low, because each stage causes losses. So there exists an *optimum for V_ϕ and N.*

- For very low voltage applications the Dickson multiplier will become very inefficient. For a supply voltage in the order of $1\,V$ the voltage increase per stage is only a few hundred mV due to V_D and the required number of stages to reach e.g. $3.3\,V$ would be about 8. In that case the last term in the expression becomes very large. In addition to much decreased efficiency there is the much increased area consumption.

For the reason of the much reduced efficiency of the Dickson Multiplier for very low supply voltages, other voltage multiplier schemes have emerged. An example of a Voltage Doubler that is bound to become a classic is treated in the next section.

3.2.3 A Voltage Doubler

The circuit of Fig. 3.6 is a voltage doubler [Fav 97] [Fav 98]. When one clock is high, the bootstrapped voltage is high enough to turn on the other nMOST, which charges the other capacitor to V_{in}. In the next phase the roles are exchanged. The great advantage of this scheme is that a fixed voltage (available during both clock phases) can be made by the application of two switches, from each internal node to V_{out}. This switch needs to be a pMOST. The problem arising with that is to ensure the reverse bias of the junctions, and it is here that the subtleties of the different designs exist.

The ratio of the lost and useful power can in a similar manner to the previous section be calculated to be

$$\frac{P_l}{P_u} = \frac{2}{V_{out}} \left(\frac{\alpha}{(1+\alpha)} V_{in}^2 \frac{C\,f}{I_{out}} + (1+\alpha) \frac{I_{out}}{C\,f} \right) \tag{3.12}$$

where α is defined as $C_s = \alpha\, C$.

Equation (3.12) is similar to (3.11) apart from the fact that there is no power loss due to diodes here. Most of the conclusions of the latter also hold for this case, with the exception of the conclusions relating to the application of diodes. In fact, the great advantage for very low voltage applications is that this voltage doubler is much more efficient than the Dickson Multiplier, by the lack of diodes in this scheme.

First Shipment (yr)	1997	1999	2001	2003	2006	2009	2012
Min. Feature Size (nm)	250	180	150	130	100	70	50
Supply Voltage (V)	1.8-2.5	1.5-1.8	1.2-1.5	1.2-1.5	0.9-1.2	0.6-0.9	0.5-0.6

Table 3.2. *The Power Supply Voltage Road-map.* [SIA 97]

3.2.4 *Advantages, Disadvantages and Limitations of the Application of Voltage Multipliers*

The use of voltage multipliers provides an easy, quick and reliable way of designing low voltage Switched Capacitor circuits. It is very similar to classic, high voltage Switched Capacitor operation, of which the design procedure is well established and well known. Furthermore it has the advantage that it cuts away an important part of the power dissipation by having the amplifiers work with the low supply voltage. For these reasons it is the choice of implementation of the industry at this moment. However, there are some disadvantages and a serious limitation as explained next.

In low voltage Switched Capacitor circuits that employ voltage multiplication to drive the switches, there is no need nor use to apply complementary switches. Usually the clock driving voltage is much higher than the opamp supply voltage and the signal range, and therefore one type of switch device can be used and only a positive voltage needs to be generated. On the other hand, in high end applications the clock driving circuit is not the same for all switches. A single voltage multiplier can create crosstalk between the nodes of the circuit in the signal path. In those cases several clock multipliers are used, each dedicated to a different group of switches [Cho 95] or possibly even dedicated to a single switch. This makes the technique even more area consuming and this is a considerable disadvantage.

There is one limiting condition to the application of voltage multipliers. It is that the used technology must be capable of withstanding the generated voltage. There are two situations to distinguish. First there are the systems with reduced supply voltage, fabricated in a "high" voltage technology. An example is [Way 92] which uses a 1 V supply in a 3 μm CMOS process. Second there are the systems designed in a low voltage technology. While the use of voltage multiplication is excellent practice for the former case, it is impossible in the latter. Table 3.2 shows the SIA CMOS technology and supply voltage road-map [SIA 97]. It shows the year of first product shipment, the minimum feature size and the corresponding power supply voltage. Because selection of power supply voltage has become part of analyses undertaken to simultaneously optimize speed and power consumption of digital IC's, it is given as a range of usable values in each technology. The trend to smaller device dimensions continues. Together with their reduced gate oxide thickness, also the maximum supply voltage of sub-micron processes scales down. In the near future deep sub-micron processes, the supply voltage will scale down to less than two Volts and eventually to less than one Volt. It is clear that in those cases - implying all systems fabricated in a deep sub-micron process - application of voltage multiplication is out of the question. This is the major drawback of this otherwise robust technique when looking at the future of Switched Capacitor techniques in general.

3.3 Multi Threshold Processes

Another option when trying to cope with reduced switched capacitor circuit supply voltage is to use devices with reduced threshold voltages. There are two ways of obtaining low V_T transistors.

The first is to use a process which besides transistors with standard threshold values, offers devices with reduced threshold voltages. The second is to use circuit techniques to reduce the threshold voltages.

3.3.1 The Use of Natural Transistors

Low-V_T transistors can be made as a by-product of a dual gate CMOS process. They can be obtained by not applying the threshold adjust implant and are for that reason called natural. The threshold voltages are given by [L&S b] [Che 90]

$$\begin{aligned} V_T &= \phi_{GB} - \frac{Q_{ox}}{C_{ox}} + 2\phi_F + \gamma \sqrt{|2\phi_F| - V_{BS}} \\ &= \pm\frac{kT}{q}\ln\frac{N_G N_B}{n_i^2} - \frac{Q_{ox}}{C_{ox}} \pm \frac{kT}{q} ln\frac{N_B}{n_i} \pm \frac{\sqrt{2q\varepsilon_{Si}N_B}}{C_{ox}} \sqrt{|2\phi_F| - V_{BS}} \end{aligned} \tag{3.13}$$

where + needs to be taken for the *n*MOST and − for the *p*MOST. The meaning of the symbols can be found in the List of Symbols on page ix. Remark that this expression is only correct for dual poly devices where the n^+ poly is used for the *n*MOST and vice versa. The threshold voltage without adjust implant is determined by the gate and bulk dopant concentration and mainly by the latter since it appears in a term which varies with the square root of it. The natural transistors' thresholds turn out to be quite symmetric for the *n* and *p* type.

The use of natural MOSTs was introduced in [Ada 90] where a 1.4V Switched Capacitor filter was implemented in a specifically modified standard CMOS technology. In [Baz 95a] natural transistors have been fabricated using a 0.5 μm CMOS process. They were measured to show $V_{T_n} \approx V_{T_p} \approx 200mV$. In [Baz 95c] a Switched Capacitor ΔΣ modulator using that technology has been fabricated, that operates with 1V supply voltage. This circuit shows the lowest supply voltage Switched Capacitor operation to date using low V_T transistors. Lower supply voltages are possible for digital circuits as in [Bur 94] where 200mV supply voltage was realized using 0V V_T devices.

The above examples used either a modified existing process and were more or less experimental. On the other hand some processes exist that are actually featuring natural transistors [Sun 92]. However, for practical implementations, the cost factor should be taken into account. Selectively shielding the threshold adjust implant is the most flexible since both high and low V_T devices are available. This an expensive operation since process steps need to be added. Removing the steps for the threshold adjust altogether for the whole wafer could even be cheaper, but is not always allowable from design standpoint. This fact makes that the application of this technique is perhaps only justifiable in special cases where production cost minimization is not an issue, such as in biomedical implementations.

3.3.2 A Circuit Technique for V_T Reduction

The MOST threshold voltage can also be reduced using a circuit technique and thus not making any changes to the physical device. Equation (3.13) reveals the dependency of V_T on the bulk to source voltage V_{BS}. Usually the bulk to source junction is reverse biased and causes a V_T increase due to the bulk effect. Reverse biasing on the other hand, causes the opposite effect, i.e. a V_T reduction. The reverse bias voltage is limited to less than 500mV due to the forward bias current that starts to flow. To date no implementations using this technique are reported.

3.3.3 Disadvantages of the Use of low V_T Transistors

Besides the threshold voltage there is another important quantity that specifies the sub-threshold behavior of a MOST. It is the sub-threshold slope. To obtain the sub-threshold slope the logarithm of the weak inversion drain current

$$I_D = I_{D0} \exp\left(\frac{V_{GS}}{n\frac{kT}{q}}\right) \tag{3.14}$$

is taken. However, both the natural and the base 10 logarithm are being used, giving

$$\frac{1}{n\frac{kT}{q}} \quad \text{for } ln \text{ representation} \tag{3.15}$$

$$\frac{1}{ln10\, n\frac{kT}{q}} \quad \text{for } log \text{ representation} \tag{3.16}$$

The latter is conceptually the most interesting. Indeed, the inverse of the sub-threshold slope is the required V_{GS} decrease in order to reduce the leakage current by a factor of ten. This could also be interpreted otherwise. Reducing the threshold voltage shifts the sub-threshold characteristic more toward the origin of the V_{GS} axis. For each decrease of V_T by an amount equal to the inverse of the sub-threshold slope, the leakage current increases by a factor of ten. A typical value for the sub-threshold slope is $80\,mV/$decade. So for a V_T reduced by $500\,mV$ compared to the standard device, this means that the leakage current is increased by a factor of $500/80 \approx 6$ orders of magnitude!

So the switch off-leakage has increased tremendously. In a Switched Capacitor integrator, the basic building block of any Switched Capacitor circuit, this leakage causes signal dependent charge loss. Any signal dependent non-ideal behavior causes harmonic distortion. So, Switched Capacitor circuits employing low V_T switches will show severe linearity degradation.

A trial in order to circumvent the leakage problem is developed in [Baz 95b]. The observation there is that a switch consisting of complementary low V_T devices connected in series have still a low on-resistance (however reduced compared to the single switch device), but shows remarkably little leakage current. The reason is that a switch in the off phase is always connected to virtual ground (e.g. $V_{DD}/2$) and that the transistor connected to that ground node always sees a large off-voltage. Of course this switch now has a reduced input range and conducts only well in the middle region, for signals that are not within V_T. This problem is solved by adding standard V_T devices in parallel to the series connection, as shown in Fig. 3.7. The added switches don't conduct in the middle region, but they do in the regions close to supply rails. Of course this solution is only possible when both the standard and low-V_T transistors are available. It further turns out that the minimum conductance has also degraded by a factor of eight compared to the only low-V_T transmission gate.

In addition to the increased leakage and distortion problem another drawback of MT processes is that the benefit is in fact not so large as one would wish. Imagine a low voltage design with NMOST only switches, having a minimum over-drive voltage of $500\,mV$. The application of a low V_T device offering a reduction of $500mV$ in threshold voltage, would cause a switch resistance reduction of only a factor of two. In fact, for a constant switch conductance, the allowable supply reduction is equal to the difference between the reduced and standard threshold, which is only a few hundred mV. Finally the circuit must be designed to cope with temperature variation. A $100°C$ temperature variation causes about 200 mV threshold voltage shift. As many circuits must be able to operate at this worst case condition, which is relatively more severe at low overdrive voltage, the benefit of using low V_T devices is even less advantageous.

3.4 The Switched Opamp Technique

From the preceding sections the following conclusions can be drawn. Voltage multipliers are the choice of implementation at this moment, because of the ease of design and reliability. However

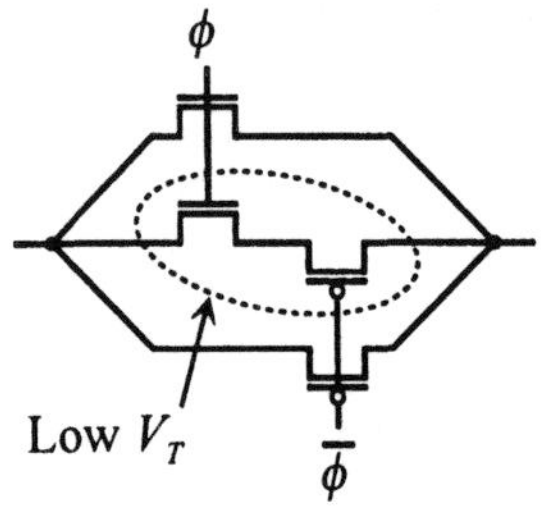

Figure 3.7. *A composite switch for application with low V_T transistors. The series connection of low V_T devices shows relatively good conductance for input signals not within V_T from the supply rails and has low leakage current. The standard V_T transistors enlarge the input range to a full rail-to-rail range.*

for high quality Switched Capacitor circuits they are area consuming and show low efficiency. The ultimate limitation toward the future is that in the future deep sub-micron technologies, which will not sustain the multiplied voltage, they will plainly not be applicable. Low V_T transistors - either the natural transistors or with forward bulk to source biasing - don't really offer a considerable switch resistance or power supply reduction, while on the contrary the off-leakage is significantly increased causing large harmonic distortion. It would be a great advantage if genuine Switched Capacitor operation could be had at low supply voltages without the disadvantages of the above discussed solutions. This is made possible by the Switched Opamp Technique.

The objectives of the development of the Switched Opamp principle were [Ste 93][Cro 94] to have genuine Switched Capacitor techniques available

- at very low supply voltages, possibly down to a single battery supply
- with low power consumption and
- the above achievable in standard CMOS technology. This consequently means the above must be done by only making use of the high V_T values available. So no low V_T transistors nor voltage multipliers will be used in the Switched Opamp Technique.

3.4.1 The Original Switched OpampPrinciple

3.4.1.1 The Key Problem.

The key problem of designing Switched Capacitor circuits below a certain power supply level is that the switches do not behave properly anymore. About $V_{Tn} + V_{Tp} + 0.5V$ is the practical minimal power supply voltage for which the switch still has rail-to-rail switch input range. Below that voltage there is a range in which the switch is not able to sample the input signal as is apparent in Fig. 3.4, and consequently, proper Switched Capacitor operation is impossible.

Looking at the Switched Capacitor integrator of Fig. 3.1 it can be concluded that there are two types of switches:

1. switches having one channel terminal connected to the reference level V_{ref}. This connection is either physical as is the case for switches S_2, S_3 and S_5 or via virtual ground as for switch S_1.
2. switches that are not connected to V_{ref} but to a signal source. These are the switch S_6 connected to the output of the amplifier, but also the switch S_4 at the very input of the system.

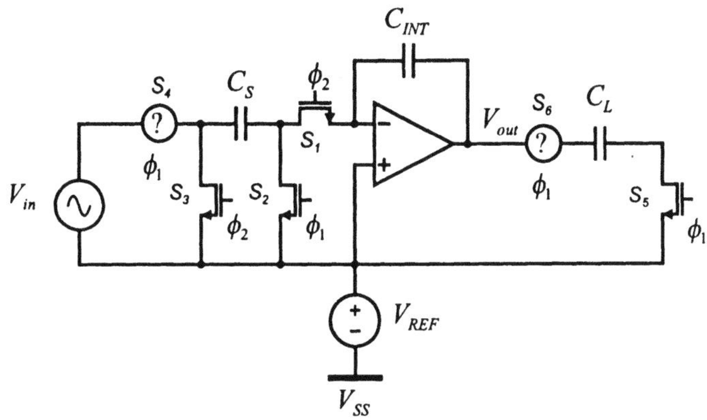

Figure 3.8. *An intermediate state in the evolution from the Switched Capacitor to the Switched Opamp integrator. The switches connected to V_{ref} can be replaced by an n-type single transistor switch. The switch at the output of the opamp and the one at the input of the system are the bottleneck.*

The first type of switch can always be turned on if the switch driving voltage is at least V_T plus an over-drive voltage V_{ov} higher than V_{ref}. So the required minimum supply voltage is

$$V_{DD} = V_{SS} + V_{ref} + V_T + V_{ov} \tag{3.17}$$

The behavior of these switches is treated in some more detail in Section 4.6.3. The second type of switch, however, needs to be able to pass the entire signal range. Because the condition under which the operation is being considered here is for a supply voltage too low to have proper complementary switch operation, these switches present the bottleneck.

An intermediate state in the evolution from the Switched Capacitor integrator of Fig. 3.1 to the Switched Opamp integrator is shown Fig. 3.8. The switches connected to V_{ref} are replaced by *n*MOST switches for the reasons put forth before in Section 3.4 on page 32. The switches at the output of the opamp and at the input of the system present the key problem. The Switched Opamp technique actually only offers a solution for (the switches at the outputs of) integrators in a chain of integrators. So for now the switch at the system input is disregarded and only the output switches are considered. Possible solutions for the problem of the input switches are treated in Section 3.4.5.

3.4.1.2 The Essence of the Switched Opamp Technique.

The Switched Opamp Technique boils down to a very simple principle, and can be formulated extremely shortly.

> *The only way to eliminate the problem of the switch at the output of an amplifier is to eliminate the switch itself.*

So the switch is plainly replaced by a short as can be seen in Fig. 3.9. This is of course easier said than done, since there are a few consequences. Special measures have to be taken to ensure proper operation.

- The obtained building block is still an integrator, but since the output of the opamp is now being sampled during the integration phase, it is a half delay integrator. If a full delay integrator is desired, a half delay unit needs to be added.

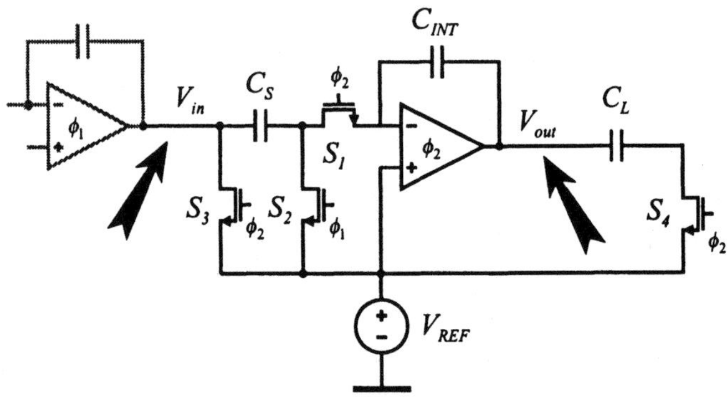

Figure 3.9. *The Switched Opamp integrator which is preceded by another integrator. The problem switches at the output of the opamps are left out, leading to a half delay integrator structure.*

- When integrating the sampled output of a preceding integrator, the output of that integrator is being shorted to V_{REF} by S_3 as can be seen in Fig. 3.9. The opamp input will see the integrator capacitor voltage drop across its inputs, and cause a large output current to flow. The latter flows through the switch to V_{REF}, but causes a voltage drop over it. This voltage drop causes inaccurate charge transfer in the next integrator, giving rise to reduced accuracy and harmonic distortion, due to the signal dependent nature of the current dumped through the switch.

  Therefore the opamp's driving force needs to be disabled. This can be done by switching off the opamp in the non-active phase. This fact has given rise to the name of the technique, i.e. Switched Opamp.

Before continuing, the time is taken to highlight an important principle to keep in mind that makes up the soul of the Switched Opamp Technique, and it is the fact that the only switches that can be used are switches connected to V_{REF}.

3.4.1.3 Half Delay Element and Full Delay Switched Opamp Integrator.

Classic Switched Capacitor integrators can be made delaying with a full or half phase delay or even not delaying. The Switched Opamp integrator obtained in the previous section incorporates by construction a half phase delay. If an integrator with a full phase delay is required, a half delay element is needed. The latter is shown in Fig. 3.10. The full delay integrator can then be constructed as a cascade of the half delay amplifier and the half delay Switched Opamp integrator. The half delay amplifier just features a gain of one by choosing C_1 and C_2 equal, and usually equal to C_S.

By construction, the integrator of Fig. 3.9 is non-inverting. It is possible to construct an inverting Switched Opamp integrator as well, but it can only be made with zero delay, which means it would be active in the same phase as the preceding integrator.

Compared to a classic Switched Capacitor non inverting delaying integrator, the Switched Opamp implementation requires two opamps. By this reason the required chip area also roughly doubles. The power consumption is not multiplied by two, however. The two amplifiers are now each active in a different phase, and they are each off during the other phase. They have a duty cycle of at most 50%, so the power consumption is in first order not affected.

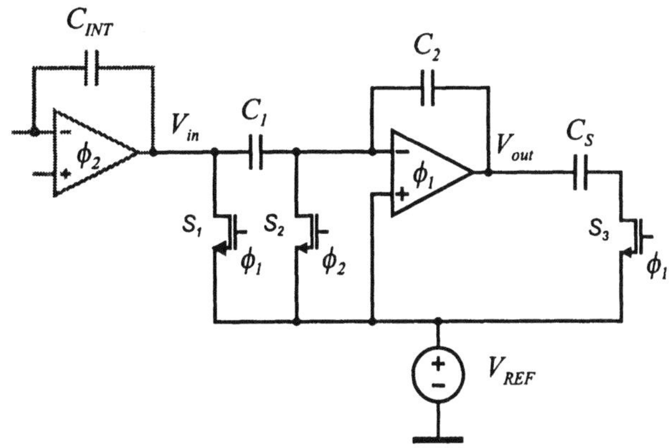

Figure 3.10. *The half delay element. In gray the part of the preceding integrator is shown.*

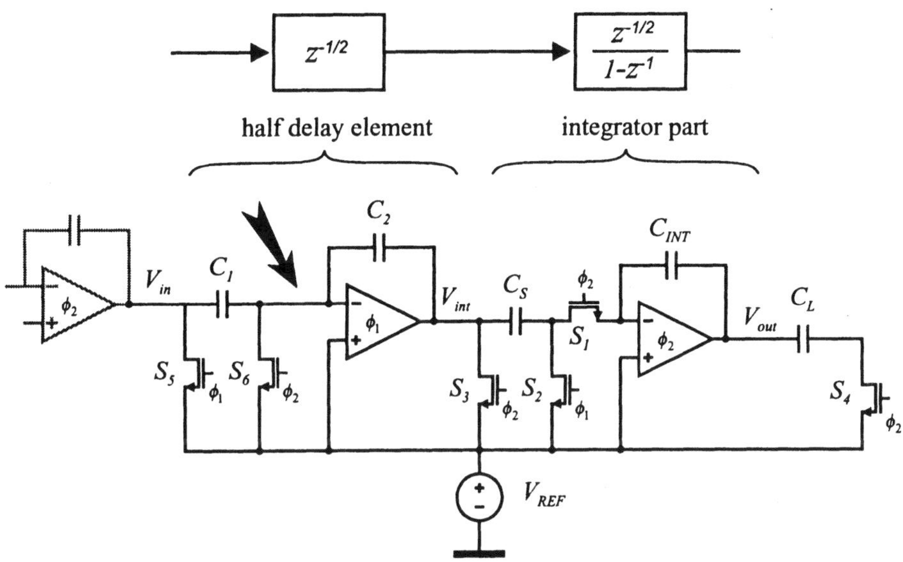

Figure 3.11. *The full delay Switched Opamp integrator consists of a cascade of a half delay element and a half delay integrator.*

3.4.1.4 About Signal Levels.

Before the principle of the Modified Switched Opamp Technique can be introduced further on, it is necessary to pay attention to some circuit issues. Fig. 3.12 shows two separate transistor networks. The one on the right is a minimal input stage. It features a differential pair consisting of $M_{IN,1}$ and $M_{IN,2}$, with a tail current source M_0, which suffices as an input stage for an amplifier to be used in a Switched Opamp integrator. The transistors between the drains of $M_{IN,1}$ and $M_{IN,2}$ and V_{SS} could be several things. They could be diode connected current mirrors or just

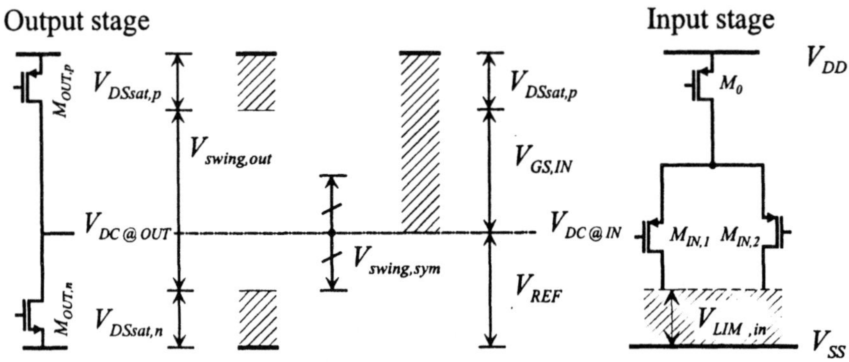

Figure 3.12. *The signal swings in the Original Switched Opamp Technique . The implicit choice is made that $V_{DC@OUT} \equiv V_{DC@IN}$. As a consequence the symmetrical signal swing is not optimal.*

current sources in case of a folded cascode branch or in case of a second amplifier stage - which is more likely in a low voltage environment. Depending on the case there they need a minimal voltage drop of $V_{LIM,in}$. On the left of the figure is the minimally imaginable output stage of an amplifier. It could be used -as is- as a second stage for the mentioned input stage. In order to be kept in saturation at all times, which is required for proper operation, $M_{OUT,p}$ and $M_{OUT,n}$ need respectively a drain to source voltage drop of $V_{DSsat,p}$ and $V_{DSsat,n}$ as is graphically indicated in the figure. This means that the range indicated by $V_{swing,out}$ is the maximal available signal swing, while the hatched zones next to the transistors of the output stage indicate the forbidden zone for the output signal. The DC level of the output swing $V_{DC@OUT}$ can be chosen freely in the available output range. In the input stage, M_0 requires also a $V_{DSsat,p}$. The input transistors $M_{IN,1}$ and $M_{IN,2}$ need a certain gate to source voltage V_{GSIN}. This fixes an upper bound on the DC signal level at the amplifiers input terminals. The amplifier's input DC voltage is at the level

$$V_{DC@IN} = V_{DD} - V_{DSsat,p} - V_{GSIN} \tag{3.18}$$

Note that all voltages are referred to V_{SS}. The amplifier's input DC voltage is further also at the reference level

$$V_{REF} = V_{DC@IN} \tag{3.19}$$

The forbidden zone for the amplifier's input DC level is indicated by the hatched region extending downward from V_{DD}. Note also that the switches, which in the Original Switched Opamp Technique are connected with one terminal to V_{REF} have a driving voltage of

$$V_{GSswitch} = V_{GS,IN} + V_{DSsat,p} \tag{3.20}$$

The question is now how to fix the DC levels. In the Original Switched Opamp Technique the implicit choice is made that

the DC level at the output of the amplifier is taken equal to the DC level at its inputs.

$$V_{DC@OUT} \equiv V_{DC@IN} \tag{3.21}$$

Consequently, the symmetrical signal swing is determined by the difference between $V_{DC@IN}$ (or the bottom of the forbidden input zone) and the closest forbidden output zone. Whether it's the

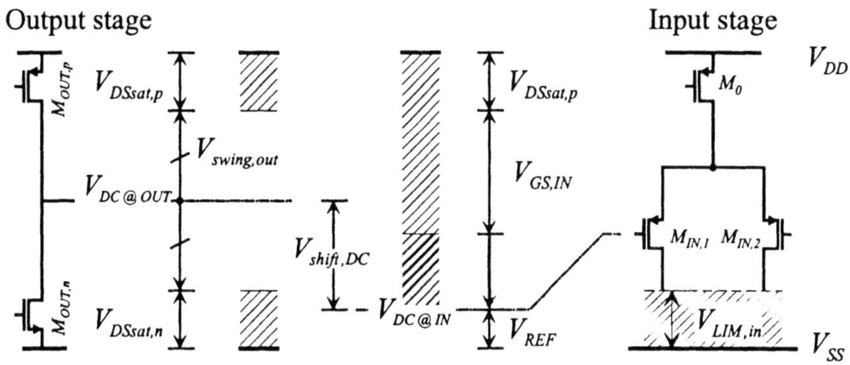

Figure 3.13. *The signal swings in the Modified Switched Opamp Technique. The explicit choice is made that $V_{DC@OUT} \neq V_{DC@IN}$. As a consequence the symmetrical signal swing is optimal and the switches have more driving voltage.*

upper or lower one depends on whether V_{REF} is lower or higher than the middle of the available output range. With the middle of the available swing being

$$V_{swing,med} = \frac{1}{2}(V_{DD} + V_{DSsat,n} - V_{DSsat,p}) \tag{3.22}$$

this is expressed as

$$V_{swing,sym} = \begin{cases} 2(V_{REF} - V_{DSsat,n}) & \text{for} \quad V_{REF} < V_{swing,med} & (a) \\ (V_{DD} - V_{DSsat,n} - V_{DSsat,p}) & \text{for} \quad V_{REF} = V_{swing,med} & (b) \\ 2(V_{DD} - V_{DSsat,p}) & \text{for} \quad V_{REF} > V_{swing,med} & (c) \end{cases} \tag{3.23}$$

Except for the singular case of (3.23)(b) the swing is not optimal. The symmetrical signal swing is smaller than the available signal swing.

$$V_{swing,sym} < V_{swing,out} \tag{3.24}$$

This is a disadvantage of the Original Switched Opamp Technique .

3.4.2 The Modified Switched Opamp Technique

The Original Switched Opamp Technique doesn't make optimally use of the available signal swing, because of an implicit choice made in fixing the DC levels of the signals. It is indisputely important to maximally exploit the available signal swing in low voltage analog circuits. The Modified Switched Opamp Technique accomplishes this, and furthermore establishes an optimal condition for the operation of the switches.

3.4.2.1 The Principle.

The framework in which the Modified Switched Opamp Technique is situated, is the same as for the Original Switched Opamp Technique . Therefore, Fig. 3.13 also shows the minimum minimorum input and output stage, respectively on the right and on the left.

The principle of the Modified Switched Opamp Technique [Bas 94] can be stated very compactly and it is that the explicit choice is made to *choose the DC level at the output of the amplifier different from to the DC level at its inputs, such that they can be separately optimized.*

This, in fact, means that the two DC levels are decoupled from each other, and that two different criteria can be used to optimize each of them. The above definition deliberately does not include the originally proposed choice of DC levels [Bas 94], such that other choices can still be considered *Modified* Switched Opamp Technique as long as they are a Switched Opamp Technique implementation with different DC levels. The originally proposed and in later described realizations implemented choices are presented next.

The DC level at the output is chosen at the middle of the output swing.

$$V_{DC@OUT} \equiv V_{swing,med} \tag{3.25}$$

In the practice of low voltage Switched Opamp circuits, the output stage will be the one depicted in Fig. 3.13, this means that it is chosen as

$$V_{DC@OUT} = \frac{V_{DD}}{2} \tag{3.26}$$

The obvious consequence and advantage over the Original Switched Opamp Technique is that now the available signal swing is optimally used as symmetrical signal swing, and it is actually a rail-to-rail swing.

$$V_{swing,sym} \equiv V_{swing,out} \tag{3.27}$$

The depicted p-type input stage's input range extends from the forbidden zone down and can in general be designed to include V_{SS}. The condition for this is stated further on. $V_{DC@IN}$ can be chosen freely in this range and this is the general case shown in Fig. 3.13. It can be chosen lower than before, such that

$$V_{DC@IN} \equiv V_{REF} < V_{DD} - V_{DSsat,p} - V_{GS,IN} \tag{3.28}$$

This puts more voltage drop over M_0 than needed, but it puts also more driving voltage over the switches, which is very welcome at low voltage operation. Therefore the Modified Switched Opamp Technique chooses the input DC level at V_{SS}

$$V_{DC@IN} \equiv V_{SS} \tag{3.29}$$

The beauty of this is that the switches receive the maximally available driving voltage of the full supply voltage:

$$V_{ov,switch} = V_{DD} \tag{3.30}$$

The condition for which the input stage's input range includes V_{SS} is

$$V_{Tp} + V_{ov,IN} > V_{DSsat,IN} + V_{LIM,in} \tag{3.31}$$

this means in practice that a normal diode type current mirror cannot be used, but a folded cascode transistor can. If a current is still wanted, special low voltage current mirrors have to be used, that take up less voltage drop than a diode connected MOST. This solution is chosen here and more details can be found in Section 4.1.2 and Section 4.2.

An important consequence of decoupling the input and output DC levels is that a voltage shift between output and input exists. In fact, the output voltage is still being sampled with respect to $V_{REF} = V_{SS}$, because sampling is only possible with reference to a level that allows

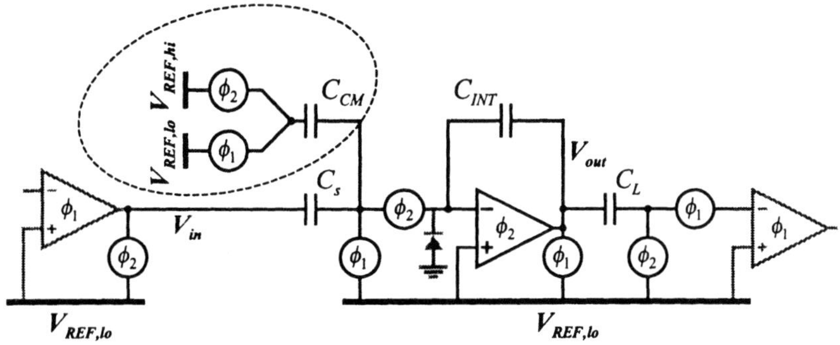

Figure 3.14. *The faulty Modified Switched Opamp integrator. C_{CM} compensates for the DC offset charge sampled on C_S. However, in this configuration the shown diode junction is in forward bias during the off phase such that the integrator charge leaks away.*

to drive a switch, which is by definition V_{REF} in the Switched Opamp Technique. The sampled output voltage has a DC offset component.

$$V_{OUT} = V_{shift,DC} + V_S \tag{3.32}$$

where the DC offset is

$$V_{shift,DC} = \frac{V_{DD}}{2} \tag{3.33}$$

This offset needs to be compensated for. The idea is to remove it with an equal charge injection of the opposite polarity that is performed with a capacitor. The originally proposed scheme [Bas 94] to do this is shown in Fig. 3.14. It shows a Switched Opamp integrator in a chain of integrators. The reference voltage is renamed $V_{REF,lo}$ here, to comply with future needs of terminology and to keep generality. $V_{REF,lo}$ is in fact equal to V_{SS}, but in a general case it doesn't have to be. Inside the dotted line ellipse, the added offset adjust capacitor C_{CM} and switches can be seen. In order to remove the DC component of C_s it is necessary to introduce a second reference level: $V_{REF,hi}$ and also the use of the *p*MOST as an additional switch device. So the levels that C_{CM} is switched to are $V_{REF,lo}$ and $V_{REF,hi}$, which is in fact V_{DD}, but in a general case it doesn't have to be, as long as there is enough headroom to drive the *p*MOST switch. The operation is as follows. During ϕ_1 C_S is sampling the output of the previous integrator and C_{CM} is emptied. During ϕ_2 the charge on C_S is transfered to C_{INT}, but C_{CM} is pulled up to $V_{REF,hi}$. So at the end of the phase C_{CM} is charged to $V_{REF,hi} - V_{REF,lo} = V_{DD}$ and has taken up charge with the opposite polarity of C_s. The proper scaling of C_{CM} is calculated in Appendix A.2 and is given by

$$C_{CM} = \frac{C_S}{2} \tag{3.34}$$

3.4.2.2 The Pitfall.

In the system just explained, unfortunately enough, there is a pitfall. It is the diffusion to bulk junction diode of the switch connected to virtual ground ('−' terminal of opamp). Assuming that $V_{REF,hi} = V_{DD}$ and $V_{REF,lo} = V_{SS}$, the following things can be said, without loosing much generality. Since the Modified Switched Opamp integrator is designed for a rail-to-rail output

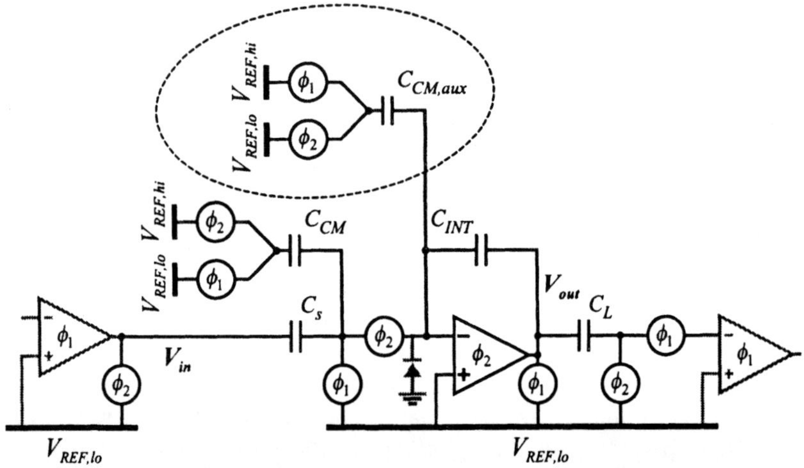

Figure 3.15. *A Modified Switched Opamp integrator cell circumventing the feed-forward and the charge loss by adding an auxiliary capacitor* $C_{CM,aux}$.

swing, the integrator output voltage can be almost V_{DD}. The other terminal of the integrating capacitor C_{INT} is at V_{SS} during the integrator phase and can therefore have a voltage drop of almost V_{DD}. The Switched Opamp operation requires that during the next phase the output of the opamp is shorted to V_{SS}. The integrator charge is trapped on C_{INT}, so when the one terminal of this capacitor goes to V_{SS}, the other terminal is pushed below V_{SS}, by an amount equal to the output voltage of the previous phase, which as stated can be almost V_{DD}. This forward biases the mentioned junction capacitor. Most of the integrator charge is therefore dumped to ground and lost and the integrator action is annihilated. It is noteworthy that it is not just a case of a transient feed-forward, but C_{INT} stays in its pushed down position during an entire half phase, which is plenty of time for the charge to flow away. It is obvious that this configuration does not allow proper functioning. A solution for this problem is required to make the Modified Switched Opamp Technique viable.

3.4.2.3 Avoiding the Pitfall.

A way to circumvent the feed-forward and integrator charge loss explained in the previous section is given in Fig. 3.15 [Cra 95]. It consists of the Modified Switched Opamp integrator cell of Fig. 3.14 plus an additional auxiliary capacitor $C_{CM,aux}$. During the integration phase ϕ_2 this capacitor's charge is transferred to the integrator capacitor C_{INT}. It thus is empty in the beginning of the off phase ϕ_1. During that phase the C_{INT} and $C_{CM,aux}$ are series connected between $V_{REF,hi}$ and $V_{REF,lo}$. The charge on C_{INT} redistributes partly to $C_{CM,aux}$. In fact, the operation is such that the middle node that is connected to the inverting opamp input always stays above $V_{REF,lo}$. The switch source to bulk junction diode will not get forward biased any longer. In the next phase ϕ_2 the charge on $C_{CM,aux}$, which is part of the intgrator charge, is transfered back to C_{INT} again. The integrator charge is temporarily stored on the auxiliary capacitor, so to speak.

The necessary condition for $C_{CM,aux}$ is derived in Appendix A.3. A sufficient and at the same time practical condition is

$$C_{CM,aux} = C_{INT} \tag{3.35}$$

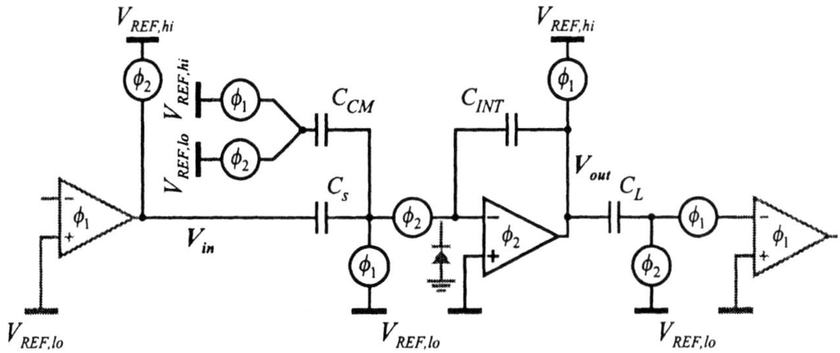

Figure 3.16. *The corrected Modified Switched Opamp integrator cell. The problem of forward biasing the source to bulk junction diode is avoided by wiring the opamp output during its off phase to $V_{REF,hi}$ instead of $V_{REF,lo}$. In that case the positive plate of C_{INT} is pulled up instead of pushed down and consequently both plates stay above V_{SS}.*

This is a substantial capacitance. C_{CM} was already $C_S/2$. With this solution another capacitor twice as large needs to be added. This is quite disadvantageous. The added capacitance increases the capacitive load by a large amount since the effective load capacitance the amplifier needs to drive is the capacitance connected at its input plus its capacitance at its output increased by the division by the feedback factor, which is smaller than unity. (cfr., Section 4.5)

$$C_{L,eff} = \left(C_S + C_{CM} + C_{CM,aux} + \frac{C_L}{1 + \frac{C_L}{C_S + C_{CM} + C_{CM,aux} + C_L}} \right) \tag{3.36}$$

Without going into detail at this moment the addition of capacitors may be a disadvantage. It either increases the power consumption for the same speed of operation, or it decreases the speed of operation for the same power consumption.

A simpler solution is introduced next. It also avoids the addition of the extra capacitor.

3.4.2.4 The Egg of Columbus.

The best solution for the problem of the junction diode that is forward biased in the off phase of the Modified Switched Opamp integrator is not far. Let's analyze the problem by means of Fig. 3.14. Capacitor C_{INT} has a large voltage drop. When the output of the opamp is shorted to $V_{REF,lo}$, this capacitor capacitively forward biases a bulk to source diode. The core of the problem is that C_{INT} is wired to $V_{REF,lo}$, by the top plate. Therefore the other plate is bound to go below $V_{REF,lo}$. If on the contrary the top plate could be pulled upwards, the bottom plate would also move up, instead of down. The solution is as obvious as the Egg of Columbus and lies in wiring the output of the opamp to $V_{REF,hi}$ instead of $V_{REF,lo}$. This reference voltage is available anyway, because it is needed for the offset charge compensation (Section 3.4.2.1). The obtained corrected Modified Switched Opamp integrator cell is shown in Fig. 3.16. [Bas 97a] [Bas 97b] In Appendix A.4 it is proven that this circuit has a normal integrator transfer function.

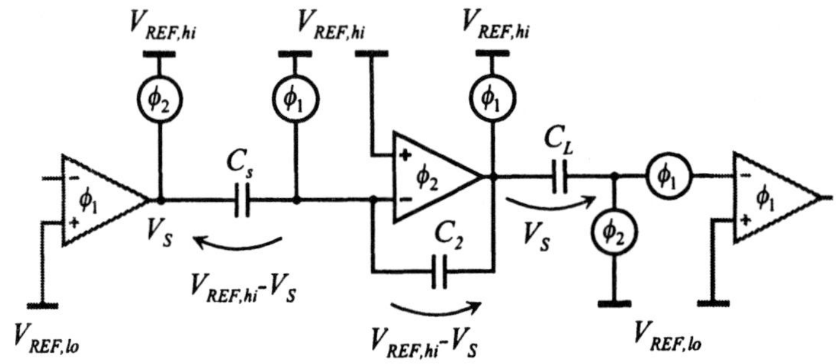

Figure 3.17. *A Minimal Modified Switched Opamp Delay Element. Sampling the signal to $V_{REF,hi}$ avoids the use of a C_{CM} and resets C_{INT} to its virtual ground node.*

3.4.2.5 A Modified Switched Opamp Delay Element.

Just as in the Original Switched Opamp Technique , it might be desirable to have a delay element available, in case its use cannot be avoided. A Modified Switched Opamp delay can be had by the addition of one switch to the Modified Switched Opamp integrator of Fig. 3.16. If a switch, active in ϕ_1, is added between $V_{REF,hi}$ and the terminal of C_{INT} that is connected to the inverting amplifier input pin, then C_{INT} is reset to zero each phase. This operation turns the integrator into an amplifier with a half phase delay. There are two disadvantages to this approach. Any parasitic capacitor connected to the inverting amplifier input is charged by a voltage difference of $V_{REF,hi} - V_{REF,lo}$. This charge introduces a DC offset. The second disadvantage is the capacitive load presented by C_{CM}. Both of these drawbacks can be avoided with the new scheme of Fig. 3.17, which is a minimal Modified Switched Opamp delay element. Now the signal to be delayed is sampled to $V_{REF,hi}$, which avoids the need of C_{CM} and the resetting of C_{INT} is done to its virtual ground point, which avoids the mentioned offset problems.

3.4.3 *Differential Modified Switched Opamp Technique*

The integrator cells introduced so far were single ended. In practice it is desirable to have a differential integrator available. There are several reasons for this. Differential building blocks have a far better Power Supply Rejection Ratio (*PSRR*) than their single ended counterparts. They have good Common Mode Rejection Ratio (*CMRR*). They also don't have even order harmonic distortion components, at least in theory, i.e. when the mismatch between the two sides is is zero. Another advantage on the system level is that they offer a sign inversion at no extra cost. It suffices to interchange the two outputs in order to get a sign inversion. On the other hand, they require a common mode feedback circuit. This topic is treated further on in Section 3.4.4.

3.4.3.1 Differential Modified Switched Opamp Integrator Cell.

It is straightforward to turn the single ended Modified Switched Opamp integrator of Fig. 3.16 into a differential version. It suffices to copy the capacitive network and apply differential amplifiers instead of single ended ones. Fig. 3.18 shows the Differential Modified Switched Opamp integrator cell. Two capacitors called $C_{CMS,eq}$ have been added. They are the equivalent common mode sensing capacitors and represent whatever capacitive network is applied to sample the common mode of the output signals, which is required to do the feedback.

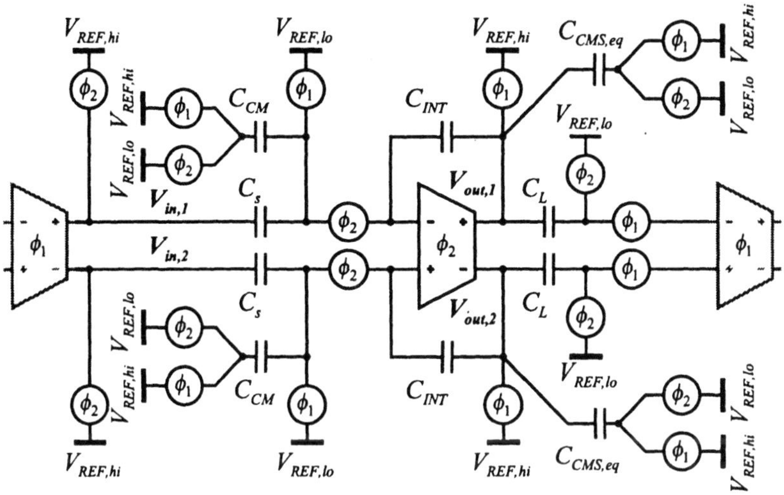

Figure 3.18. *The Differential Modified Switched Opamp integrator cell. It is derived from Fig. 3.16 by doubling the capacitive network and applying differential amplifiers. Capacitors* $C_{CMS,eq}$ *are equievalent capacitors representing the common mode sampling network load from Fig. 3.20, Fig. 3.21 or Fig. 3.22.*

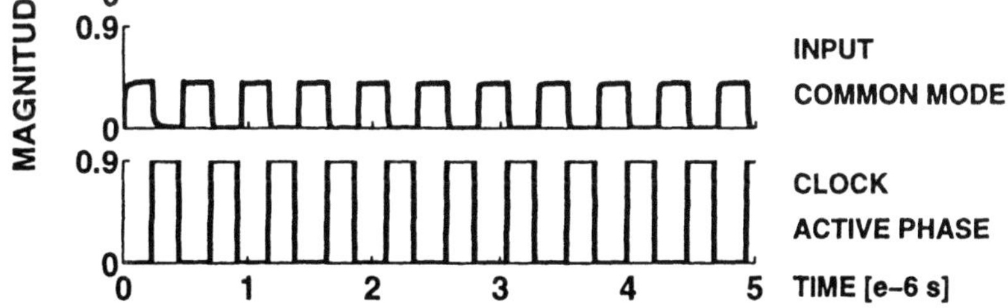

Figure 3.19. *A transient simulation of the Differential Modified Switched Opamp integrator for a supply voltage of 900 mV. The top curve is the differential output and shows the integrating action for a constant input. The other curves show the output and input common mode signals and the applied clock. The active clock phase is* ϕ_2 *in Fig. 3.18.*

Fig. 3.19 shows a transistor level transient simulation of the Differential Modified Switched Opamp integrator. The amplifier used is the novel low voltage class AB OTA presented in Section 4.4. The supply voltage is 900 mV and the period of the applied clock signal (shown in the bottom graph) is 470ns. The applied differential input is a constant value of 40 mV. The integrator action can be seen in the top graph which shows the differential output. The graph below shows the output common mode. During the active phase, it is in the middle of the output signal range at half the supply voltage, and it is pulled up to V_{DD} in the off phase as explained in the above sections. The third graph from the top is the common mode of the input signals. In the on phase it is at V_{SS} as required by the Modified Switched Opamp Technique and in the off phase it is pulled upwards to stay above V_{SS}.

3.4.4 *Common Mode Feedback Principles in the Differential Modified Switched Opamp Technique*

Fully differential circuits require a control loop to regulate the common mode of the signals. This system is generally called a Common Mode Feed Back system (CMFB). While the signal processing is only differential, the CMFB makes sure that the signals are properly positioned in the range where they can be processed by the circuit. In practice this means that the CMFB keeps the common mode of the output signals in the middle of the available output swing.

3.4.4.1 Why Classic Systems Fail.

Many CMFB circuits exist. For Switched Capacitor circuits the most popular is the so called Switched Capacitor CMFB. [Cal 89][Cas 85][Cas 91] It cannot be used in the Switched Opamp methodology. The main problem is that a passing type switch is needed, connected to the output of the amplifier. As explained in Section 3.4.1.2 the only switches that can be used in the Switched Opamp Technique are switches connected to the reference levels, $V_{REF,lo}$ and $V_{REF,hi}$.

3.4.4.2 An Error Amplifier Based Approach.

The CMFB problem in the Switched Opamp Technique is to sample the common mode output signal *and* at the same time cope with the low supply voltage. Sampling the output common mode shouldn't be too difficult. A capacitive divider, which is reset each phase can do the job. However, the difficulty is in applying the sample to the input of the CMFB amplifier. The latter in general does not have a rail-to-rail input stage. The input range will be small and close to one supply rail, if based on a differential pair input stage, while the common mode sample taken is more or less in the middle between the two supply rails.

A CMFB scheme especially designed for the Switched Opamp Technique is shown in Fig. 3.20. It is based on [Cra 95], but adapted to the *corrected* Modified Switched Opamp integrator of Fig. 3.16 [Pel 98b][Pel 98a], where the output is wired to $V_{REF,hi}$ in the off phase, instead of to $V_{REF,lo}$ as it was originally proposed. It is based on a capacitive sampler and divider. The operation is as follows. The inputs to the capacitive network in_1 and in_2 are connected to the outputs $V_{out,1}$ and $V_{out,2}$ of the Modified Switched Opamp Integrator of Fig. 3.18. During ϕ_1, the opamp is off and those outputs are wired to $V_{REF,hi}$. The other terminals of the capacitors are also shorted to $V_{REF,hi}$ and consequently all capacitors are reset to zero. During ϕ_2, when the opamp is on, the capacitive network consisting of C_{CMS2} and two instances of C_{CMS} samples the output common mode voltage and at the same time divides the voltage difference between the common mode sample and $V_{REF,lo}$. The capacitive division is established by the addition of C_{CMS2}. The reason for this division is to be able to apply the taken sample to the input of an error amplifier that does not have a rail-to-rail input. It in general allows a more beneficial biasing of the input stage, i.e., with a larger V_{GS}. The division ratio α is given by the ratio of the

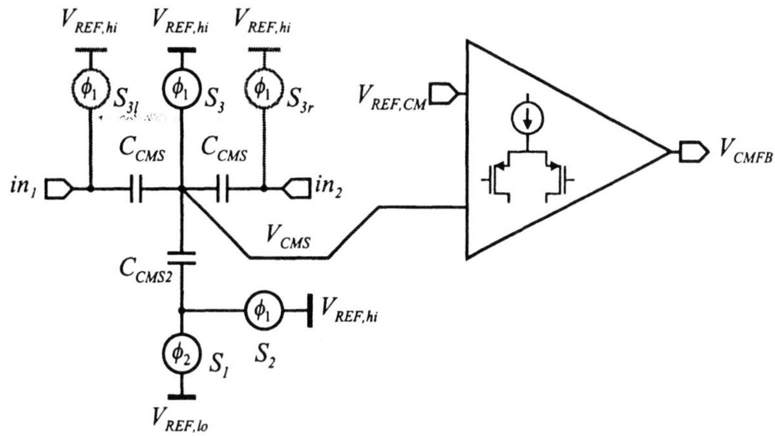

Figure 3.20. *An error amplifier based CMFB. Shown is on the left a capacitive voltage CM sampler and divider and on the right an error amplifier. By doubling the capacitive network and applying differential amplifiers.*

sampled common mode voltage V_{CMS} and the real common mode voltage V_{CM} and it is given by

$$\alpha = \frac{V_{CMS}}{V_{CM}} = \frac{1}{1 + \frac{C_{CMS2}}{2\,C_{CMS}}} \tag{3.37}$$

and can be freely chosen. However, it has a strong influence on the speed of the CMFB settling process. The division factor α appears to be at the same time the feedback factor of the CMFB control loop. Consequently it significantly enlarges the effective capacitive load. This matter is discussed in more detail in Section 4.5.3

3.4.4.3 An Integrating Control Loop Based approach.

A second proposed common mode feedback for the Differential Modified Switched Opamp technique is shown in Fig. 3.21 [Bas 97a][Bas 97c][Bas 97b]. It does not make use of an error amplifier, but of an integrating control loop. The integrator output voltage is at the output of the differential OTA. For figure clarity the capacitors associated with the differential mode are not shown. The capacitors C_s take a sample of the output common mode voltage in the active phase ϕ_2. The capacitor C_{CM} is preset with a sign inverted charge corresponding to the desired output common mode voltage. The difference between the measured and preset common mode charge is integrated on C_A. This changes the CM feedback signal on terminal CMFB, which feeds back to the common mode control input of the differential amplifier. The proper size for C_{CM} can be found to be

$$C_{CM} = C_S \tag{3.38}$$

The advantage of this method is that the input terminals of the CMFB amplifier can be at $V_{REF,lo}$.

3.4.4.4 An Error Amplifier Based CMFB without Division.

The error amplifier based approach with the capacitive divider of Section 3.4.4.2 has one significant disadvantage, and it is that the division factor is also the feedback factor. Consequently

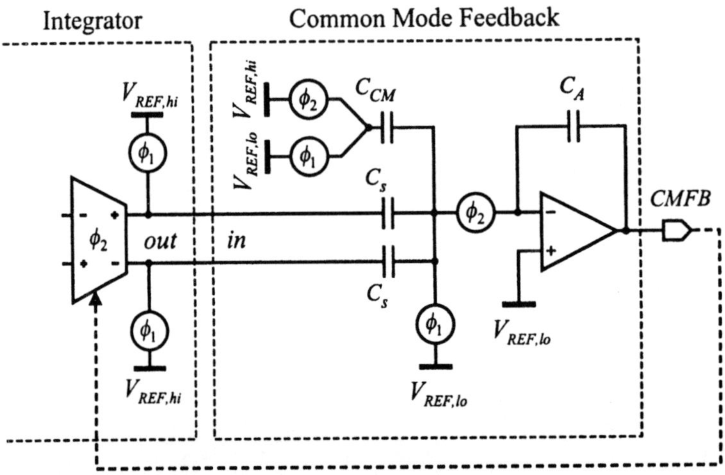

Figure 3.21. *An integrating control loop based CMFB. A common mode error injects a net charge into C_A, which stores the CMFB signal.*

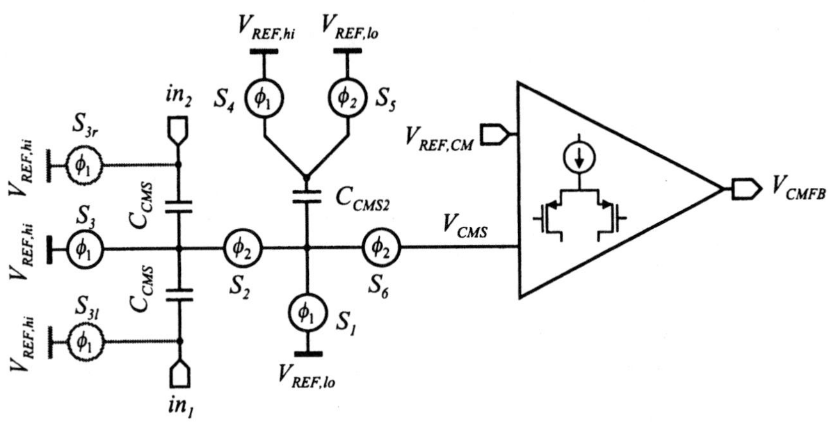

Figure 3.22. *An error amplifier based CMFB without divider. The capacitors C_{CMS} sample the common mode, while C_{CMS2} gives a DC offset to level shift the CM sample.*

it causes the effective capacitive load of the CMFB loop to be substantially increased. One may wish to keep the simplicity of this approach without this disadvantage. And this is possible as follows.

Instead of dividing the common mode sample it is possible to just give it an offset. It suffices to change the switching scheme to the one given in Fig. 3.22. The capacitors C_{CMS} now take the sample while C_{CMS2} injects charge in the middle node that effectuates an offset. The proper size in order to level shift the sample to $V_{REF,lo}$ is

$$C_{CMS2} = C_{CMS} \tag{3.39}$$

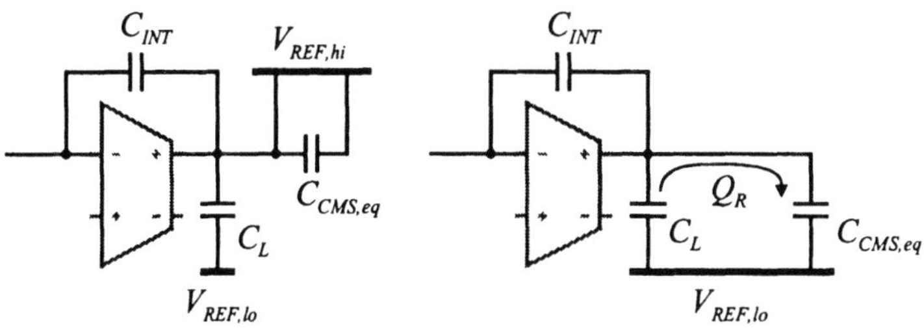

Figure 3.23. *An important advantage of the Error Amplifier based CMFB is that a charge redistribution takes place, which establishes the correct common mode voltage without action of the CMFB amplifier.*

There still exists a feedback factor now, but it is larger

$$\mathcal{F}_{dc} = \frac{2C_{CMS}}{2C_{CMS} + C_{CMS2}} = \frac{2}{3} \tag{3.40}$$

As a consequence the effective CMFB capacitive load remains much smaller than in the previous case.

3.4.4.5 Advantages of the Error Amplifier based CMFB.

By analyzing Fig. 3.18 it is clear that during the integration phase of the integrator for which C_L is the input capacitor, C_L ends the phase fully charged to $V_{REF,hi} - V_{REF,lo}$. Hence C_L starts the integration phase of the integrator where it is sampling the output fully charged. In other words, at that point in time C_L has an incorrect common mode voltage. The CMFB must then remove half of its charge to establish the correct common mode value. This process requires time. In the error amplifier based CMFB approaches, it happens so that the equivalent common mode sample capacitors $C_{CMs,eq}$, start the integration phase completely discharged. In order for them to have the correct common mode voltage, they need to be charged. In fact, because C_L is fully charged and $C_{CMs,eq}$ is discharged and they are switched to a parallel configuration, an immediate redistribution takes place between them. This process is shown in Fig. 3.23. It can be ensured that the final voltage of the redistribution equals the correct common mode value, if the following condition is met:

$$C_{CMs,eq} = C_{L,t} \tag{3.41}$$

where $C_{L,t}$ is the total load capacitance at the output of the opamp. It is important to consider the total load capacitance, because there is a large parasitic capacitor present at the output node. It is the parasitic bottom plate capacitor of C_{INT}. The great advantage is thus that an initial redistribution to the correct common mode value happens, without having to add extra capacitors. By exploiting this fact, it is possible to relax the speed requirement of the CMFB amplifier. Such a redistribution can of course also be obtained with the integrating control CMFB system, but only by adding extra capacitance, which increases the load of both the integrator opamp and the CMFB.

3.4.5 *The Switched Opamp Input Problem and Solutions*

As mentioned in Section 3.4.1.1 a sampling switch cannot be avoided at the input of a Switched Opamp circuit. There are a few options to overcome this problem, which are compatible with

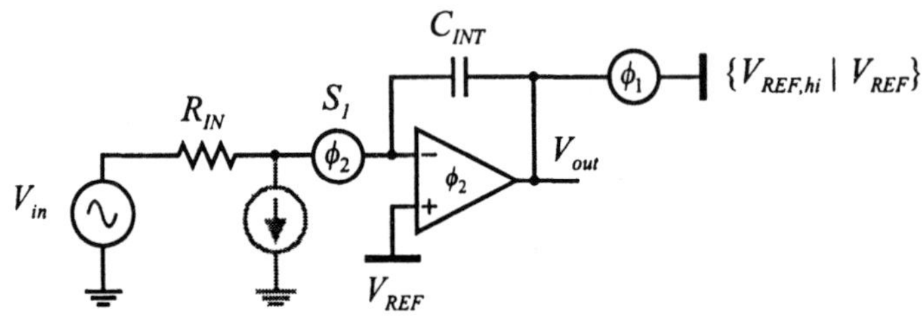

Figure 3.24. *A resistor can be used in the input section of a Switched Opamp circuit. A level shift in the integrator, generated with a constant current (current source in grey) avoids the need of a rail-to-rail input stage.*

the Switched Opamp goals. One exists of using a resistor in the input. Another one, which has been developed in this work optimally exploits the available input range.

3.4.5.1 Using a Physical Resistor at the Input.

Reflecting about the fact that a switched capacitor behaves as a resistor, it is not far-fetched to replace the simulated resistor by a physical one at the input of the Switched Capacitor circuit, as in Fig. 3.24. In this case one could speak of a switched resistor. The switch between the resistor and the integrating capacitor is needed to prevent leakage of C_{INT}. The resistor should have a value of

$$R = \frac{1}{f_s C} \tag{3.42}$$

where f_S is the sampling frequency. For f_S equal to 1 MHz and a 1 pF capacitor this requires a resistance of 1 $M\Omega$. This is very large resistor to implement on chip. Apart from the active area taken up, there is even a more stringent limitation. It is the frequency limitation due to the distributed RC time constant. Further, if a rail-to-rail swing is desired, the input common mode or DC voltage needs to be approximately in the middle of the power supplies. This causes a problem of biasing the input stage. A way to circumvent that is to apply a level shift to the input signal by generating a voltage drop over the input resistor with a current source, indicated in grey in the figure. This current must be supplied by the signal source at the input or by another current source at that position. Since the current source at the input terminal of the amplifier needs a certain saturation voltage of a few hundred mV, it is not possible to use $V_{REF,lo} = V_{SS}$. Adding current sources unfortunately also means adding noise sources right at the input of the system, which is not desirable.

Another approach using resistors to tackle the input problem is to precede the first integrator with a buffer. Since the first integrator is not preceded by an OTA, the Switched Opamp Technique cannot be applied there. Preceding it with a switched buffer can be a solution. A new type of switched buffer uses resistive feedback and a Switched Capacitor generated DC level shift between the input DC level and the input pin of the OTA which sits at V_{SS} level. [Bas 98]

3.4.5.2 A Low-Voltage Solution.

In this book a low voltage solution for the input sampling scheme is developed for the Original Switched Opamp Technique [Pel 97c]. It is shown in Fig. 3.25. The reader is reminded that the *Original* Switched Opamp Technique only deploys one reference voltage V_{REF}, situated in between V_{SS} and V_{DD}. The basic idea is that the switches are designed to be able to switch to V_{REF}. This is the maximum signal that can be passed by the switch. If the input signal is applied

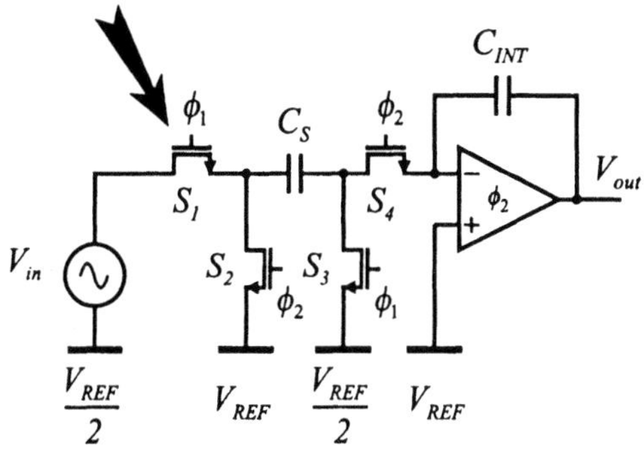

Figure 3.25. *A low voltage* Original *Switched Opamp input section, which maximally exploits the available swings.*

symmetrically around $V_{REF}/2$, the maximum input signal amplitude, can be $V_{REF}/2$ as well. It is still integrated with reference to V_{REF} . Fig. 3.26 shows the signal levels for the conditions $V_{DD} = 1.5V$, $V_{REF} = 0.5V$ and $V_{GS,switch} = 1V$. The available input swing is thus

$$V_{swing,in} = V_{REF} \tag{3.43}$$

In this text the major interest is the design of $\Delta\Sigma$ modulators. Because the internal signal distribution in such a system is always within the range $[-V_{REF,\Delta\Sigma}, +V_{REF,\Delta\Sigma}]$ the $\Delta\Sigma$ reference voltage can be chosen as high as half of the symmetrical opamp output swing, here denoted by the internal signal swing $V_{swing,int}$ which is available on the nodes *internal* to the Switched Opamp filter.

$$V_{REF,\Delta\Sigma} = \frac{V_{swing,int}}{2} \tag{3.44}$$

For instance, the stable input range of a second order $\Delta\Sigma$ modulator is about $0.75V_{REF,\Delta\Sigma}$ [Pel 97a][Mar 98b]. It follows that the available input swing for the structure of Fig. 3.25 is smaller than the maximal modulator input range. To make full use of the dynamic range of the 2nd order modulator, the input signal should be scaled. In order to bring the maximum input signal up to the maximum modulator input level a multiplication factor should be built in defined by

$$F = \frac{2\gamma V_{REF,\Delta\Sigma}}{V_{swing,in}} \tag{3.45}$$

where γ is the stable input range relative to $V_{REF,\Delta\Sigma}$. This factor is applied by proper scaling of C_s.

The multiplication factor F that is built into the first stage is not always needed. For a 3rd or 4th order single loop $\Delta\Sigma$ modulator the stable input range is $[-0.5V_{REF,\Delta\Sigma}, +0.5V_{REF,\Delta\Sigma}]$ (making $\gamma = 0.5$) while the internal signal distribution is between $[-V_{REF,\Delta\Sigma}, +V_{REF,\Delta\Sigma}]$. When the OTA output swing is about a factor of two larger than the switcheable input swing, combination of the two above equations shows that $F = 1$, so no scaling of the input needs to be carried out.

3.5 Conclusion

In this chapter the key problem of low voltage Switched Capacitor circuits is posed. Existing solutions are presented. From these the Switched Opamp Technique is selected for further ex-

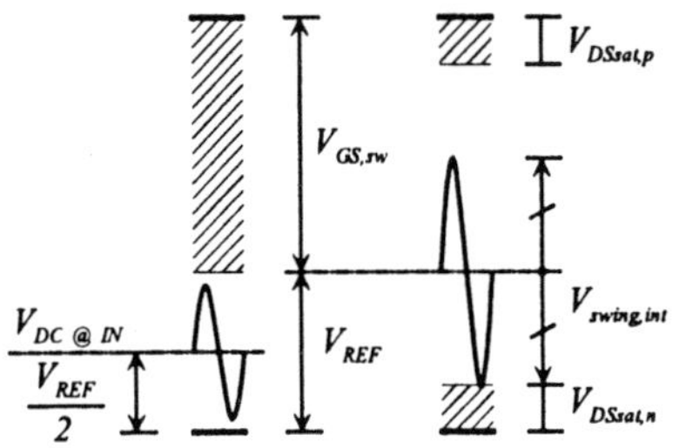

(a) *Original Switched Opamp*

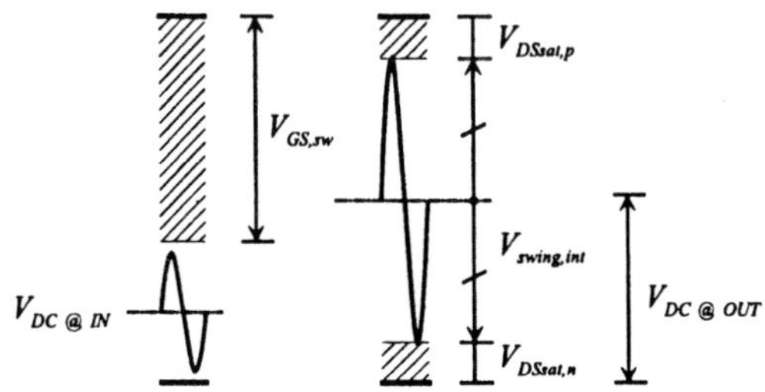

(b) *Modified Switched Opamp*

Figure 3.26. *The signal swings at the Switched Opamp input and internally. (a) For the original Switched Opamp Technique where the (internal) symmetrical swing is smaller than the available amplifier output swing and (b) for the Modified Switched Opamp Technique where it is optimal.*

ploration, and the existing building blocks are developed, in a logical and more or less historical sequence.

The core problem of low voltage signal processing is the switch driving problem. Turning on the switches becomes impossible below a certain supply voltage. Voltage multipliers provide a quick and easy way to overcome this problem. Their drawbacks are limited efficiency causing increased power consumption, and the impossibility of application in future deep sub-micron CMOS processes. Special technologies providing low V_T transistors to be used as switches are costly, not widely available and suffer from off-leakage, while the benefit is not all that large. The Switched Opamp Technique is capable of performing real low voltage Switched Capacitor signal processing, without use of voltage multiplication nor low V_T transistors. The main idea is to leave out the switch at the output of the amplifier, which is the real bottleneck at very low supply voltage. The Original Switched Opamp Technique is limited in supply voltage reduction due to an implicit choice made which is to take the DC levels at the input and output of the integrator equal. The Modified Switched Opamp Technique decouples these levels and is capable of achieving lower supply voltage while providing optimal switch over-drive voltage. A differential integrator cell based on the Modified Switched Opamp Technique is presented. Several CMFB schemes are presented. At the input of the first Switched Opamp integrator a sampling switch cannot be avoided. This switch is also a bottleneck. A solution to overcome this, making optimally use of the available input and internal swings is presented.

4
LOW VOLTAGE CIRCUIT DESIGN

In Chapter 3 the core problem of low voltage Switched Capacitor circuits is tackled: it is the switch driving problem. At that point, in Section 3.4.1.1, abstraction is made of the feasibility of operating amplifiers at low voltage. This chapter shows how to achieve low voltage operation on the circuit level. The design techniques are demonstrated with the two OTAs that are used in the realizations of Section 6.1 and Section 6.2. The first structure is a two stage Miller OTA and the second one is a new low voltage class AB OTA especially designed for Differential Switched Opamp applications. Further, other low voltage design techniques are demonstrated for the required building blocks in $\Delta\Sigma$ modulators.

4.1 Towards Low Supply voltage

In this section the power supply limits in the design of opamps are scanned. At the same time a comparison of the minimal power supply for the Original Switched Opamp Technique and Modified Switched Opamp Technique can be made. A few words are written about current mirrors and cascode transistors.

4.1.1 Minimal Supply Voltage

Some issues in the design of low voltage opamps have already been touched upon in Section 3.4.1.4 and Section 3.4.2.1, where that input was the driving force for the development of the Modified Switched Opamp Technique. Opamps can be designed with quite low power supply. Consider Fig. 3.12 on page 44. There a minimal input stage (based on a differential pair) and a minimal rail-to-rail output stage section are shown. These can be combined in several ways to achieve an OTA with rail-to-rail output stage. The first is a cascade connection, the second is with a current mirror and the third is by adding a folded cascode branch plus a

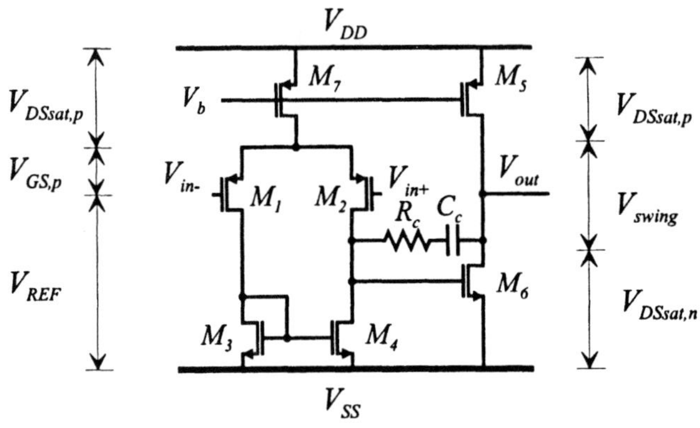

Figure 4.1. *The two stage Miller OTA.*

cascade connection. The former of these yields e.g., the two stage Miller OTA of Fig. 4.1. It is single ended, so it uses a current mirror in the input stage and a cascade connection of the single output stage. Some elements restricting its input and output range are shown. This is the opamp which is proposed for the Original Switched Opamp Technique [Cro 94] and used in the realization of Section 6.2.

The minimum supply voltage for this OTA and at the same time for the Original Switched Opamp Technique can be expressed as

$$V_{DDOSOT,min} = V_{DSsat,n} + V_{DSsat,p} + V_{GSIN} + \frac{V_{swing,sym}}{2} \tag{4.1}$$

This is under the condition that

$$V_{REF} < V_{swing,med} \tag{4.2}$$

and

$$\begin{aligned} V_{LIM} &< V_{REF} + V_{GSIN} - V_{DSsat,IN} \\ &\lesssim V_{REF} + V_{T,IN} \end{aligned} \tag{4.3}$$

where $V_{swing,med}$ is the middle of the available output swing and V_{LIM} is the limiting voltage over whatever structure is to be used at the drains of the input stage. The reader is reminded that in the *Original* Switched Opamp Technique there is only one reference voltage V_{REF}. There is no use in the opposite of condition (4.2) because that situation provides less switch overdrive voltage. Therefore it is not considered. If (4.2) is satisfied, the reference voltage for the Original Switched Opamp case is

$$V_{REF} = \frac{V_{swing,sym}}{2} + V_{DSsat,n} \tag{4.4}$$

A simplified expression for the minimum V_{DD} is

$$V_{DDOSOT,min} = 2\,V_{DSsat} + V_{T,IN} + V_{ov,IN} + \frac{V_{swing,sym}}{2} \tag{4.5}$$

The minimum bound on the power supply can be seen from (4.5) to be limited by the saturation voltages of the differential pair tail current mirror, and of the bottom output transistors, the threshold voltage of the input transistors and their over-drive voltage and finally the desired swing. The required power supply is thus a function of the desired swing. Taking V_{DSsat} 200 mV, V_T 700 mV, $V_{ov,IN}$ 200 mV and a swing of 400 mV a minimum supply of 1.5V is achievable with this topology. If the input transistors are biased in the transition region between weak and strong inversion, the swing can be enlarged to 600 mV while keeping the same supply voltage.

The quantity V_{LIM} of Fig. 3.12 puts a limit on the lower bound of V_{REF} via the saturation voltage and V_{GS} of the input devices as expressed by (4.3). V_{Lim} can be any of the following:

V_{GS}	for a simple current mirror
V_{DSsat}	for a folded cascode
V_{GSn}	for a cascade connection with an nMOST stage
$V_{DD} - V_{GSp}$	for a cascade connection with a pMOST stage.

In the case of the single ended two stage Miller OTA, the first and third cases apply. Condition (4.3) is easily satisfied.

Reducing the supply voltage of the two stage Miller OTA or of the Original Switched Opamp Technique for that matter even further can only go at the cost of reduced signal swing. Lower supply voltage operation than obtained with the above approach is however possible, without sacrificing signal swing. The Modified Switched Opamp Technique of Section 3.4.2.1 provides the methodology. The signal levels are visualized in Fig. 3.13. The key element there is to take V_{REF} zero. The minimal power supply can for the Modified Switched Opamp Technique be seen to be

$$V_{DD_{MOSOT,min}} = min\left\{V_{DSsat} + V_{GSIN}, 2\,V_{DSsat} + V_{swing,sym}\right\} \tag{4.6}$$

From this expression it can be seen that the required swing may become a more stringent limitation on the minimal supply voltage of the Modified Switched Opamp Technique than the threshold voltage. Comparing (4.6) with (4.5) it can be concluded that a lower supply voltage is possible or that for the same supply voltage a larger swing is obtained both with the Modified Switched Opamp Technique.

With the constraint of zero V_{REF}, the condition of (4.3) is not met anymore when a simple current mirror or an nMOST second stage is deployed. Possible solutions for differential topologies are to use a pMOST second stage or even a folded cascode branch with an nMOST second stage.

At supply voltages reaching the order of 1 V the application of the simple current mirror becomes problematic. Having a low voltage current mirror available would increase the flexibility of designing low voltage amplifiers. For that reason a low voltage current mirror is discussed in the next section.

4.1.2 Current Mirrors

Current mirrors are important elementary transistor stages. In this section the low voltage performance of several current mirror topologies is discussed. In order to do this, an input and output voltage drop are defined. They are called ΔV_{IN} and ΔV_{OUT}. They are the minimum required voltage drop needed for proper biasing and operation, between the input respectively output terminal and the supply rail they are connected to.

The simple current mirror is shown in Fig. 4.2(a). It requires an input voltage drop of V_{GS} and an output voltage drop of V_{DSsat}. Often a cascoded current mirror is desired for high output impedance and for better current matching, as shown in Fig. 4.2(b). This configuration requires an input voltage drop of not less than $2V_{GS}$ and $V_{GS} + V_{DSsat}$ at the output. It is obvious that this cascoded current mirror is not meant for low voltage applications. A much improved cascoded

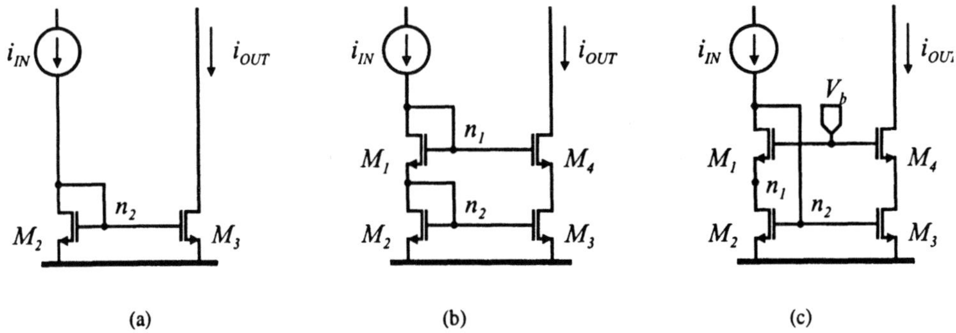

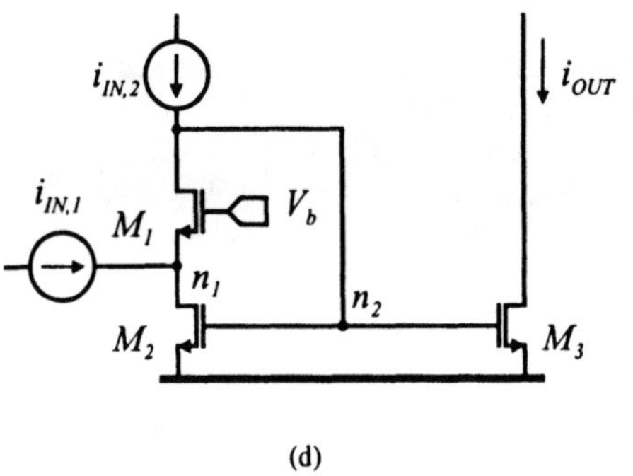

Figure 4.2. *Different types of current mirrors.*
(a) Simple current mirror, (b) Cascoded current mirror, (c) Cascoded current mirror with increased output swing and (d) True low voltage current mirror.

current mirror is shown in Fig. 4.2(c) [Fis 87][Bab 88] also known as the wide output swing cascode current mirror. It requires an input voltage drop equal to the simple current mirror and has an output voltage drop of $2V_{DSsat}$,. So, not a large penalty is paid for the benefit of the cascode feature, compared to the simple current mirror. Yet, it is desirable to reduce the input voltage drop even more. The scheme of Fig. 4.2(d) accomplishes this.

The topology shown in Fig. 4.2(d) is a system with feedback [Rij 93] [Kin 97] . If only $i_{IN,2}$ is considered it is identical to the wide output swing cascode current mirror of Fig. 4.2(c). Here that current is considered to be a bias and n_1 is considered as the input node. The input signal is the current $i_{IN,1}$. The input device is M_1, which is connected in a cascode configuration. The input terminal is its source and the output is its drain. The output signal is a voltage. The voltage is measured and fed back to the input as a current through device M_2. Consequently it is a s-shunt feedback amplifier. Because of the feedback loop, the actual amplifier only has to process the error signal, which is very small. That means that M_2 must draw virtually the entire input current $i_{IN,1}$. Since the gate of M_2 is connected to M_3, that current is also drawn by the latter transistor. This circuit thus buffers or mirrors the input current. This operation can be more intuitively be described as follows. Input current $i_{IN,1}$ has the tendency to flow into the low impedance input presented to it by the source of M_1. The current emerges at the drain node and generates a large voltage swing, because it is dumped into a high impedance node. This high voltage is applied at the gate of M_2, and it generates a large current to draw $i_{IN,1}$ downward. The remaining error current is a factor of the loop gain smaller than the input current.

It must not be forgotten that $i_{IN,2}$ also can be used as an input. The output current is thus the sum (or the difference depending on the sign of $i_{IN,1}$) of the two input currents. The output current will be

$$i_{OUT} = B(i_{IN,1} + i_{IN,2}), \tag{4.7}$$

where B is the ratio of the aspect ratios of M_3 and M_2. Currents can of course also be summed or subtracted at the input node itself, a property that is made used of in the new low voltage class AB OTA of Section 4.4.

This low voltage current mirror has some important properties that need to be investigated, such as the frequency behavior, stability since it is a feedback system and input impedance. These aspects are studied in Section 4.2.

Table 4.1 summarizes the input and output voltage drops required by each current mirror of Fig. 4.2. Note that the low voltage mirror of (d) just introduced can also be cascoded as the one of (c) if desired. Also noted at this point is the fact that the low voltage operation of the current mirror comes at the cost of an extra current branch, causing more current consumption.

	(a)	(b)	(c)	(d)
ΔV_{IN}	V_{GS}	$2V_{GS}$	V_{GS}	V_{DSsat}
ΔV_{OUT}	V_{DSsat}	$V_{GS} + V_{DSsat}$	$2V_{DSsat}$	V_{DSsat}

Table 4.1. *Summary of the required input (ΔV_{IN}) and output (ΔV_{OUT}) voltage drops required by the current mirrors of Fig. 4.2.*

4.1.3 Cascoding

It has not been said with so many words in the first section, but application of cascode transistors to increase the gain is very limited in low voltage design. The reason is that it takes up a V_{DSsat}, of voltage drop. There are mainly two different cases to consider. The first is the output stage and the second is the input stage. In the first case the minimum power supply would be increased with V_{DSsat}, in the case of the Original Switched Opamp Technique and $2V_{DSsat}$, in the case of the Modified Switched Opamp Technique. This puts a very large restriction on the signal swing when the supply voltages considered are in the order of $1\,V$. Therefore cascaded (or multi stage) structures are resorted to, when high gain is required. On the other hand cascoding has been suggested as a means of achieving lower supply voltage in the first section of this chapter. A folded cascode transistor can be applied in the input stage. It makes the $V_{LIM,in}$ of Fig. 3.12 and Fig. 3.13 equal to a V_{DSsat}, instead of V_{GS} of the second stage transistor and alleviates thus the downward bottleneck for the input stage. It sounds contradictory that a cascode can allow lower supply voltage, but it isn't. Firstly, it is only the case for a folded cascode. Secondly, it can only be applied because the swing on the output node of the first stage (with folded cascode output stage) is small. This swing is small because it equals the output swing divided by the gain of the second stage

$$V_{swing,int} = \frac{V_{swing,out}}{A_2} \tag{4.8}$$

where A_2 is the gain of the second stage. The internal swing $V_{swing,int}$ at the folded cascode output of the first stage is thus only a few tens of mV. Note that in this case the achievement of low voltage by applying the folded cascode, has required the addition of an extra branch, drawing a bias current and thus consuming extra power that is only partly compensated by the obtained voltage reduction.

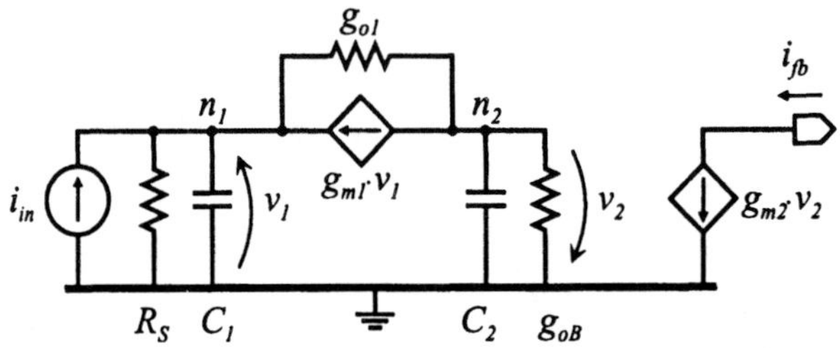

Figure 4.3. *The small signal equivalent circuit of the low voltage current mirror of Fig. 4.2(d) in open loop configuration.*

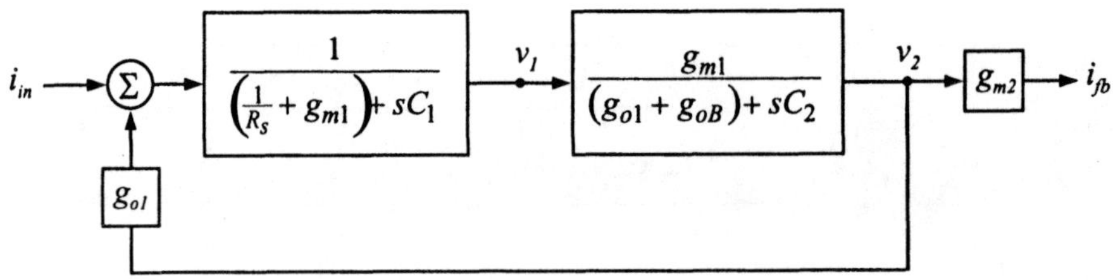

Figure 4.4. *The block diagram corresponding to the small signal model of Fig. 4.3.*

4.2 Analysis of the Low Voltage Current Mirror

In this section the important properties of the low voltage current mirror are studied, such as the frequency behavior and impedances. In order to carry out the study, the loop of the circuit in Fig. 4.2(d) is opened at the source of M_1 and the corresponding small signal equivalent circuit is drawn. It is given in Fig. 4.3.

The nodal equations yield

$$\left\{\begin{array}{lcl} v_1 & = & \frac{1}{(\frac{1}{R_S}+g_{o1}+g_{m1})+sC_1}(i_{in}+g_{o1}v_2) \\ v_2 & = & v_1\left(\frac{g_{o1}+g_{m1}}{(g_{o1}+g_{oB})+sC_2}\right) \\ i_{fb} & = & g_{m2}v_2 \end{array}\right. \tag{4.9}$$

In these equations the following definitions hold:

$$g_{oB} = \text{the output conductance of the top biasing current source}$$

$$C_1 = C_{GS1}+C_{SB1}+C_{DB2}+C_{IN} \tag{4.10}$$

$$C_2 = C_{GS2}+C_{GS3}+C_{DB1}+C_B \quad \text{,where} \tag{4.11}$$

$$C_{IN} = \text{the capacitance of the input current source, usually } C_{DB} \tag{4.12}$$

g_{o1} can be neglected compared to g_{m1}. The corresponding block diagram is shown in Fig. 4.4. Further in the analysis here where an ideal current input source $i_{IN,1}$ is considered, R_S is taken infinity. The open loop transfer function is the ratio of the feedback current i_{fb} by the input current i_{in} and can be calculated to be

$$A(s) = \frac{i_{fb}}{i_{in}} = \frac{\frac{g_{m2}}{g_{oB}}}{\left(1 + s\left(\frac{C_2}{g_{oB}} + \frac{C_1(g_{o1}+g_{oB})}{g_{m1}g_{o1}}\right) + s^2\frac{C_1C_2}{g_{m1}g_{oB}}\right)} \tag{4.13}$$

$$\approx \frac{\frac{g_{m2}}{g_{oB}}}{\left(1 + s\frac{C_2}{g_{oB}}\right)\left(1 + s\frac{C_1}{g_{m1}}\right)}$$

It can be seen that there are two poles, occurring at the two nodes of the circuit. The dominant one is on what can be considered the output node of the amplifier that is going to be fed back. The non dominant is on its input node.

Closing the loop with unity gain feedback, the obtained transfer function from input to output is found to be

$$H(s) = \frac{i_{out}}{i_{in}} \approx -\frac{g_{m3}}{g_{m2}} \frac{1}{\left(1 + s\frac{C_2}{g_{m2}} + s^2\frac{C_1C_2}{g_{m1}g_{m2}}\right)} \tag{4.14}$$

In unity feedback configuration the open loop poles shift to closed loop poles that are located much closer together and potentially cause a complex pole pair. A two-pole feedback system is absolutely stable [L&S c], but the locations of the poles are important for phase margin and transient or settling characteristics. The properties are studied by the loop-gain bandwidth product ω_{GBW} and the open loop non dominant pole.

$$A_{OLp} \approx \frac{g_{m2}}{g_{oB}} \tag{4.15}$$

$$\omega_{GBW} \approx A_{OLp}\beta\omega_{OLp,d} \approx \frac{g_{m2}}{C_2} \tag{4.16}$$

$$\omega_{nd} \approx \frac{g_{m1}}{C_1} \tag{4.17}$$

For $\omega_{OLp,nd} > 4\omega_{CLp,d}$ the poles are real and a large phase margin results. When $\omega_{OLp,nd} < 4\omega_{CLp,d}$ the poles are complex. It is common practice to design a continuous time feedback system for an open loop non-dominant pole to closed loop dominant pole ratio of 3, which corresponds to a phase margin of about 72° or a damping factor of 0.87.

The above expressions can be further simplified by distinguishing between the dominant and non dominant terms. If the parasitic capacitors are neglected in (4.10) and (4.11) and M_2 and M_3 are assumed to be equal in size, they reduce to

$$C_1 \approx C_{GS1} \tag{4.18}$$

$$C_2 \approx 2C_{GS2} \tag{4.19}$$

The obtained simplified poles are summarized in Table 4.2. The gain-bandwidth approaches $f_{T,2}/2$ while the non dominant pole approaches $f_{T,1}$. The f_T of a MOST device is given by

$$f_T = \frac{g_m}{C_{GS}} \approx \frac{1}{2\pi}\frac{3\mu(V_{GS} - V_T)}{2L^2} \tag{4.20}$$

Assuming equal length transistors and equal V_{GST1} and V_{GST2} and not necessarily equal currents, the ratio of the non dominant pole and the gain-bandwidth is approximately equal to two. This means that by nature a reasonable phase margin of 66° is obtained, which corresponds to a damping factor of 0.7. In practice, however, all the different design parameters (V_{GST}, L, I) and the parasitic capacitances can strongly alter this ratio. This is even more so in very low voltage design, where transistors are often biased closer to or in the weak inversion regime and the parasitic capacitances become increasingly more important than in the strong inversion regime.

R_S	ω_{GBW}	ω_{nd}
∞	$\frac{g_{m2}}{C_2} \approx \frac{g_{m2}}{2C_{GS2}}$	$\frac{g_{mI}}{C_I} \approx \frac{g_{mI}}{C_{GSI}}$
$1/g_{mI}$	$\frac{g_{m2}}{2C_2} \approx \frac{g_{m2}}{4C_{GS2}}$	$\frac{2g_{mI}}{C_I} \approx \frac{g_{mI}}{C_{GSI}}$

Table 4.2. *Summary and simplified expressions of poles for low voltage current mirror and new input stage.*

The feedback has a beneficial effect on the input impedance of the low voltage current mirror. The input impedance is divided by the open loop transfer function. Without feedback the input impedance is $(g_{mI} + sC_I)$. Since not the whole input current, but only the error current flows into the source of M_I after feedback, the input current generates a much smaller voltage signal on the input node. So, the feedback has decreased the input impedance, which is now given as

$$\begin{aligned} z_{in} &= \frac{1}{(g_{mI} + sC_I)(1 + A(s))} \\ &\approx \frac{1}{g_{mI}A_{OLp}} \frac{\left(1 + s\frac{C_2}{g_{oB}}\right)}{\left(1 + s\frac{C_2}{g_{m2}} + s^2\left(\frac{C_2C_I}{g_{m2}g_{mI}}\right)\right)} \end{aligned} \tag{4.21}$$

This means that for low frequencies up until the open loop dominant pole, the input impedance is very low and the input can be considered to be a virtually fixed node. This property has caused the spark of inspiration necessary for the new building block introduced in the next section.

4.3 A Differential Input Stage

In this section a differential input stage, based on the low voltage current mirror/buffer circuit is introduced. Subsequently the properties, including frequency behavior and voltage requirements are studied.

4.3.1 *The Differential Input Stage.*

The low voltage current mirror circuit studied in the previous section is steered with high impedance current source. It has the interesting property of a very low input impedance. This means that -also for large signals- the input node can be considered virtually fixed in voltage. In the current mirror configuration V_b is used to fix the voltage level at the current input node. If an extra transistor is added and connected to the input of the current mirror with its source,

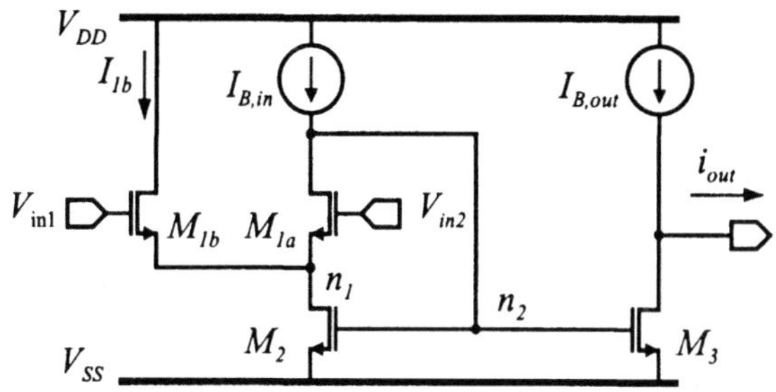

Figure 4.5. *A differential input structure based on the low voltage current mirror of Fig. 4.2(d).*

that source can be considered fixed. If a voltage signal is applied at the input of the added transistor, a current is generated and injected in the current mirror, and mirrored to the output. The resulting circuit is shown in Fig. 4.5. The added transistor is called M_{1b}. It is not necessary to bias M_{1a} with a constant voltage: its gate can be used as an extra input (here V_{in2}). The current feedback loop keeps the current constant through M_{1a}, which acts as a source follower. Consequently, if input signals V_{in2} and V_{in1} are applied, node n_1 undergoes a voltage excursion, equal to V_{in2}. Now v_{GS1b} equals $V_{in1} - V_{in2}$, so a larger current is generated through M_{1b}. In practice differential inputs are applied. Node n_1 therefore sees the opposite excursion such that the full differential input signal is applied between the gate and source terminals. In addition to i_{out}, the drain current of M_{1b} can be considered an output current as well. Hence two tranfer functions must be studied in Fig. 4.6:

$$H_{out} = \frac{i_{out}}{v_{in}} \tag{4.22}$$

$$H_{1b} = \frac{i_{1b}}{v_{in}} \tag{4.23}$$

These transfer functions are discussed in the following sections

4.3.2 Frequency Behavior of $\boldsymbol{H_{out}}$

Here the frequency characteristics of the input stage are studied. Fig. 4.6 shows the small signal model of the circuit in Fig. 4.5. The nodal equations for n_1, n_2 and the output node are

$$\left\{\begin{array}{lclcl} 0 & = & g_{m1}(v_{in}/2 - v_1) + g_{m1}(-v_{in}/2 - v_1) - v_1 Y_1 - g_{m2}v_2 & & \\ v_2 & = & -i_2 Z_2 & = & -g_{m1}(-v_{in}/2 - v_1)Z_2 \\ i_{out} & = & -g_{m3}v_2 = -g_{m2}v_2 & & \end{array}\right. \tag{4.24}$$

In these the following definitions hold

$$\begin{array}{rcl} Y_1 & = & g_{o2} + sC_1 \approx sC_1 \\ Z_2 & = & (g_{oB} + sC_2)^{-1} \\ C_1 & = & C_{GS1a} + C_{GS1b} + C_{SB1a} + C_{SB1b} + C_{DB2} \\ C_2 & = & C_{GS2} + C_{GS3} + C_{DB1a} + C_B \\ g_{m3} & = & g_{m2} \end{array} \tag{4.25}$$

Note that g_{m3} is assumed equal to g_{m2}. Capacitance C_{GS} is to be taken in C_1 . If there should be any doubt about the latter, simply rewriting the first nodal equation with C_{GS} separately will

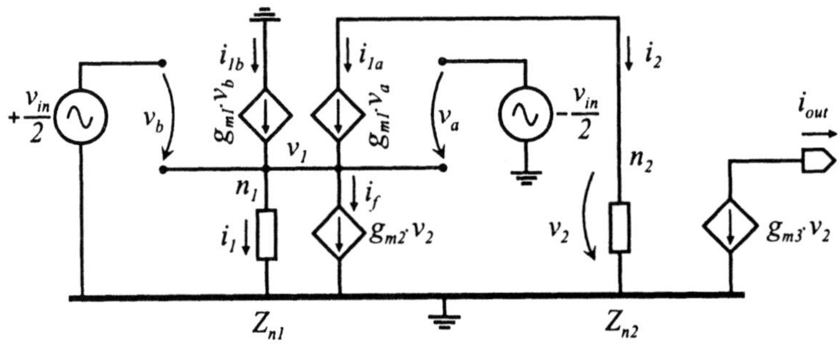

Figure 4.6. *Small signal model of differential input structure of Fig. 4.5*

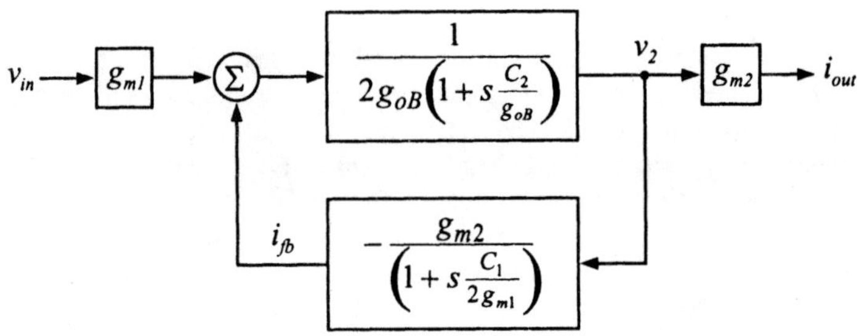

Figure 4.7. *Block diagram of small signal model of Fig. 4.6.*

readily reveal that the appear only as a sum. g_{o2} proves to be negligible compared to $2g_{m1}$, which is the impedance level seen at n_1. The nodal equations yield

$$\begin{cases} v_1 = \dfrac{-g_{m2}}{2g_{m1}+sC_1} v_2 \\ v_2 = \dfrac{g_{m2}}{(g_{oB}+sC_2)} \left(\dfrac{v_{in}}{2}+v_1\right) \\ i_{out} = -g_{m2} v_2 \end{cases} \tag{4.26}$$

Fig. 4.7 shows the corresponding block diagram. The operation is as follows. The input signal generates a current through g_{m1}, which is dumped onto n_1. The feedback current from g_{m2} is also dumped onto n_1 and summed to the input current. The feedback current has a negative sign, so it is actually subtracted from the input current. Only half of the difference of the currents (error current) reaches n_2. This is because the current dumped into n_1 is split between the the two impedance branches it sees, each with impedance $1/g_{m1}$. On n_2 the injected current generates a voltage in g_{oB}. The frequency limitation on the two nodes is apparent: a pole equal to the node impedance by the node capacitance exists on each node. Node n_2 is a high impedance (g_{oB}) while n_1 is a low impedance ($1/2g_{m1}$), so the former is the dominant pole and the latter the non dominant one.

The overall transfer function is given by

$$H_{in\rightarrow out}(s) = \frac{i_{out}}{v_{in}} \approx g_{m1} \frac{\left(1+s\dfrac{C_1}{2g_{m1}}\right)}{\left(1+s\dfrac{2C_2}{g_{m2}}+s^2\dfrac{C_2C_1}{g_{m2}g_{m1}}\right)} \tag{4.27}$$

Again, as for the low voltage current mirror, this is a second order feedback system, of which the properties are studied by the loop-gain bandwidth product and the open loop non dominant pole. Contrary to the current mirror, now a zero is also present.

$$A_{OLp} \approx \frac{g_{m2}}{2g_{oB}} \tag{4.28}$$

$$\omega_{GBW} \approx A_{OLp}\beta\omega_{OLp,d} \approx \frac{g_{m2}}{2C_2} \tag{4.29}$$

$$\omega_{nd} \approx \frac{2g_{mI}}{C_I} \tag{4.30}$$

$$\omega_{CL,z} \approx \frac{2g_{mI}}{C_I} \tag{4.31}$$

Notice that these expressions are different from the ones of the current mirror. The loop gain is a factor of two smaller. This is caused by the fact that half of the feedback current undergoes a current division by a factor of two. Two low impedance branches are present at n_I, but only one is in the loop. The other current is plainly lost. This makes the *GBW* a factor of two smaller. The open loop non dominant pole on the other hand seems to be increased because the node impedance is halved by the presence of the second branch. However, the node capacitance is also increased. The simplified expressions of Table 4.2 reveal a halved *GBW* and the same non dominant pole. Now their ratio equals 4, so by nature this system is approximately critically damped. Again, these are just very rudimentary interpretations. The detailed design depends much on the surrounding circuitry.

It will be of interest to the practical design procedure of the class AB OTA of Section 4.4 to consider a special condition for the location of the poles. The exact expression for the poles is

$$s_{1,2} = -\frac{g_{mI}}{C_I}\left(1 \pm \sqrt{1 - \frac{g_{m2}C_I}{C_2 g_{mI}}}\right) \tag{4.32}$$

For $\frac{g_{mI}}{C_I} > \frac{g_{m2}}{C_2}$ (or $\omega_{nd} > 4\omega_{GBW}$) the poles are real. Now, under the condition that $\omega_{nd} >> 4\omega_{GBW}$ the poles can be simplified to

$$\frac{g_{mI}}{C_I} >> \frac{g_{m2}}{C_2} \quad \Longrightarrow \quad \begin{cases} s_1 = -\frac{g_{m2}}{C_2} \\ s_2 = -\frac{2g_{mI}}{C_I} \end{cases} \tag{4.33}$$

In this case one pole cancels with the zero and the transfer function simplifies to

$$H_{in \to out}(s) \approx g_{mI}\frac{1}{\left(1 + s\frac{2C_2}{g_{m2}}\right)} \tag{4.34}$$

The behavior for this condition when the pole on n_I occurs at much higher frequency than the *GBW* of the loop, can be interpreted as follows. At frequencies higher than the latter there is no feedback anymore in the loop and all transistors draw a constant bias current. At that point the input transistors M_{Ia} and M_{Ib} act as a differential pair. Their sources become AC ground and consequently no poles or zeros can be associated with it any longer. A signal current flows, but not into M_2 anymore, but in M_{Ia} and M_{Ib}. The current through M_{Ia} is dumped into n_2, which at that frequency can be considered shorted to ground as well, because the dominant pole is associated with it.

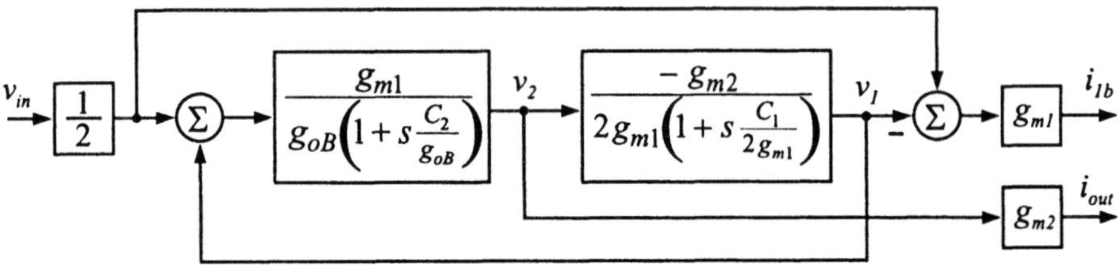

Figure 4.8. *Block diagram of small signal model of Fig. 4.6 for determination of i_{lb}.*

4.3.3 Frequency Behavior of H_{1b}

From Fig. 4.6 the nodal equations can be derived. They are the same as (4.24), except that now the current of interest is

$$i_{lb} = g_{ml}\left(\frac{v_{in}}{2} - v_l\right) \tag{4.35}$$

The block diagram is now more appropriately drawn as in Fig. 4.8. The transfer function from v_{in} to i_{lb} is now given as

$$H_{in\rightarrow 1b}(s) = \frac{i_{lb}}{v_{in}} \approx g_{ml}\frac{\left(1+s\frac{C_2}{g_{m2}}+s^2\frac{C_2C_l}{2g_{m2}g_{ml}}\right)}{\left(1+s\frac{2C_2}{g_{m2}}+s^2\frac{C_2C_l}{g_{m2}g_{ml}}\right)} \tag{4.36}$$

Again, under the condition that $\omega_{nd} >> 4\omega_{GBW}$ this result can be simplified to

$$\frac{g_{ml}}{C_l} >> \frac{g_{m2}}{C_2} \quad \Longrightarrow \quad H_{in\rightarrow 1b}(s) = \frac{i_{lb}}{v_{in}} \approx g_{ml}\frac{\left(1+s\frac{C_2}{g_{m2}}\right)}{\left(1+s\frac{2C_2}{g_{m2}}\right)} \tag{4.37}$$

For low input signal frequencies the full differential input is applied across M_{lb} and the current i_{lb} is equal to $g_{ml}v_{in}$. At frequencies higher than the *GBW*, the input transistors act as a differential pair as explained in the previous section. For very high frequencies, the current i_{lb} is then $g_{ml}v_{in}/2$. Consequently a zero at the double of the pole frequency must accompany the pole. [L&S d]

The discussed differential input stage and the low voltage current mirror can be considered the cornerstones of a low voltage class AB OTA, which is presented next.

4.4 A class AB OTA

Class A circuits are defined as circuits where the transistors are conducting current throughout the entire signal waveform. A property is that the current swing is smaller than the bias current. These circuits have quite low power efficiency, because of the power dissipation of the quiescent current. Class B circuits improve the efficiency by operating transistors at zero quiescent current. They have the disadvantage of a dead zone, i.e., in open loop configuration there is no output current when the input is below a certain threshold voltage. For this reason they are not suitable for application in Switched Capacitor integrators. Complete charge transfer is not possible if the output current vanishes below a certain input voltage. This problem is remedied by class AB circuits. These maintain the benefits of class B circuits, without having a dead zone,

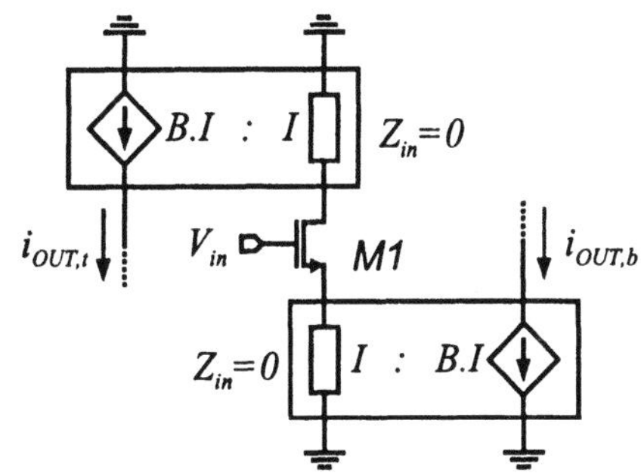

Figure 4.9. *The principle of operation of the low voltage class AB OTA.*

by biasing the transistors with a relatively low quiescent current level. Consequently it would be beneficial for low power consumption if the Switched Opamp Technique could be combined with class AB operation of the opamps. Since the output current is proportional to the square of the input voltage or even an exponential function of it (depending on the region of operation) rather than a linear function, the settling can be faster. Adversely, for a similar settling speed, quiescent current can be reduced, resulting in less power consumption.

4.4.1 The Topology

In order to construct a class AB OTA for use in the Modified Switched Opamp Technique, the following strategy is folowed. We want to:

- generate a large output current with applied input voltage
- recuperate the current at the drain and the source of the input transistor for push-pull purposes
- have input common mode range include V_{SS} or V_{DD}
- have rail-to-rail output swing
- accomplish low supply voltage operation

A single transistor accomplishes the first requirement. When the input transistor's source terminal is fixed, an input signal generates a drain current that is a quadratic or exponential function of it. Tying the source to ground doesn't allow to re-use the source current. In order to recuperate that, a building block with three particular properties is needed: (a) it should have very low input impedance as to draw all the current injected at the input node and to keep the source of the input transistor fixed. (b) it should at the same time mirror the input current to a different branch. (c) it should have good low voltage properties. That building block could then also be used to mirror the drain current to another branch. The principal circuit of this operation is shown in Fig. 4.9. A building block with the mentioned three properties is the low voltage current mirror of Fig. 4.2(d) introduced in Section 4.1.2 and analyzed in Section 4.2. In fact this addition of a transistor to the low voltage current mirror has given rise to the differential input stage of Fig. 4.5 introduced in Section 4.3. The input range of the latter includes V_{DD} if it is an *n*MOST input stage and V_{SS} if it is *p*MOST. Fig. 4.10 shows the transistor circuit corresponding to the principle model, with a *p*MOST input transistor. Instead of having just one input, the second available input can and must be used. If it is not used, but biased with a constant voltage, the input DC level offsets alter the biasing currents and thus also the settling characteristics, which behavior is not allowable. If $v_{in1} - v_{in2}$ is positive, a large current is generated, and mirrored to either branch, of which one is sourcing and the other is sinking current. When

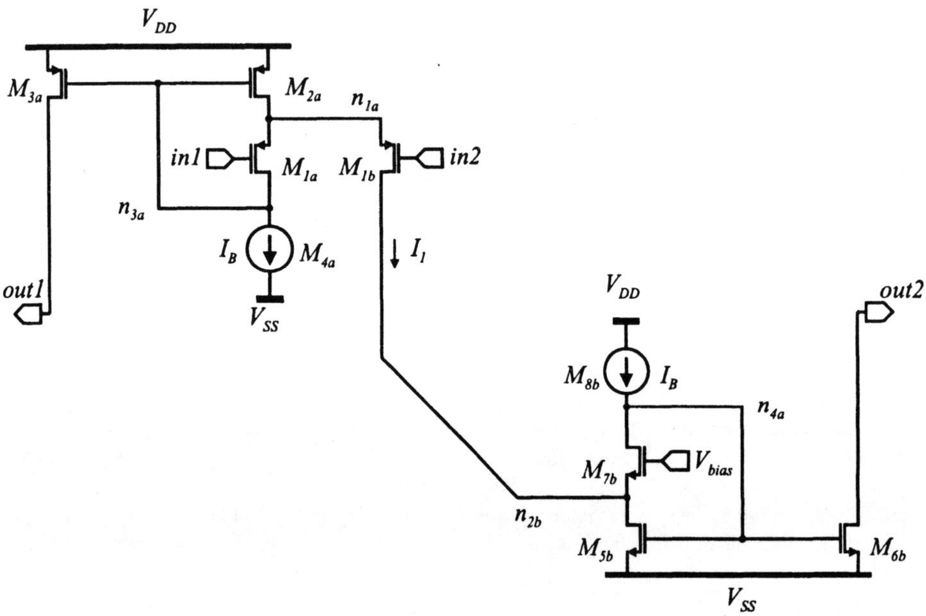

Figure 4.10. *Transistor circuit of the half class AB OTA with pMOST input stage.*

the opposite input is applied, the input transistor is pinched off, and the output currents drop to zero. In order to have fully differential class AB operation, in this case also a sourcing and a sinking current must be generated in the output branches. This can easily be accomplished by copying and mirroring the circuit already present in Fig. 4.10. Doing so gives rise to the fully differential class AB OTA and it is shown in Fig. 4.11. [Pel 98b][Pel 98a]

4.4.2 DC Transfer Characteristics

The transfer characteristic of this OTA is very typical for a class AB amplifier. It is shown in Fig. 4.12 for a certain sizing of the transistors and for differential input voltages between -500 to +500 mV. The output current is linear with small input signals, for large inputs a large current is drawn and finally the output current saturates. In this example all branches draw the same quiescent bias current of 1 μA. The ratio of the maximum single ended output current and the biasing current of one input transistor is as large as 120.

First the input output relation is derived with a simple model for operation in strong inversion. Then a more detailed analysis is given, which fits more to the reality of the specific design of the realization of Section 6.2.

4.4.2.1 A Simple Large Signal Model.

Consider the circuit in Fig. 4.5. The input transistors M_{1i} are supposed to be biased in strong inversion. The quiescent gate source voltage drop is then defined by

$$I_B = \beta_1 V_{GST1,Q}^{\ 2} \tag{4.38}$$

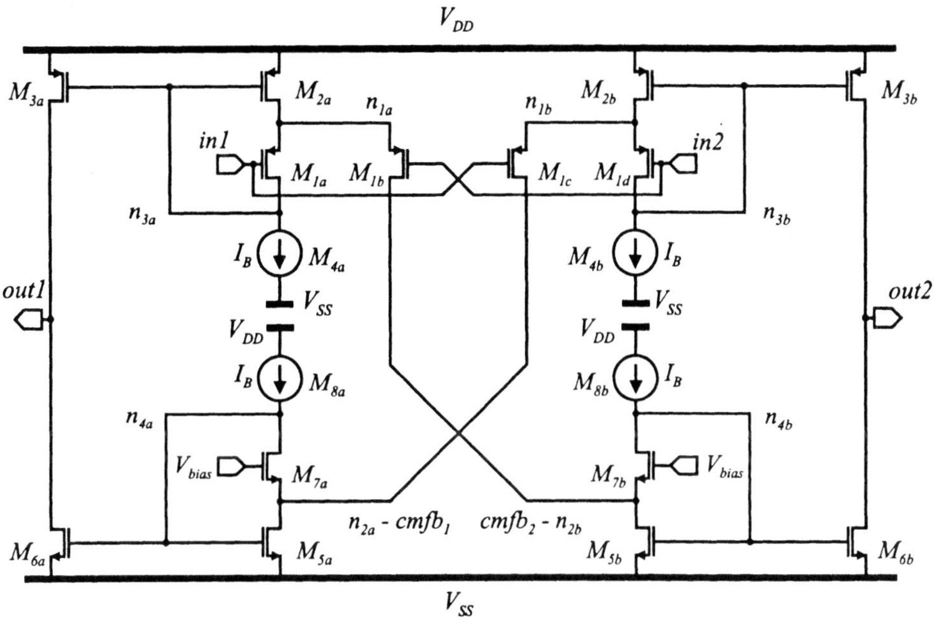

Figure 4.11. *The fully differential low voltage class AB OTA.*

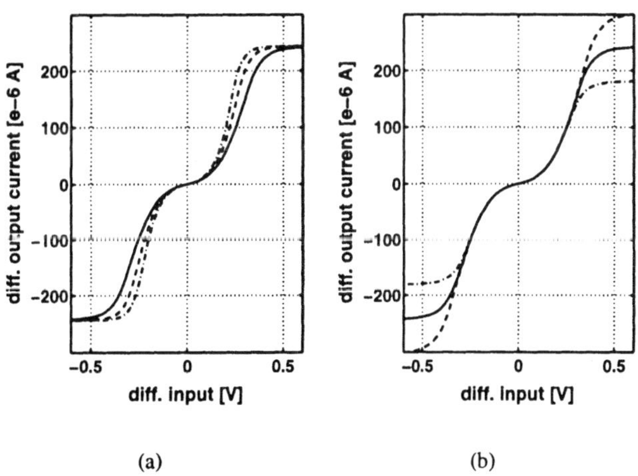

Figure 4.12. *The class AB OTA transfer characteristic. Fig. (a) shows that wider input transistors increase the steepness of the curve and (b) shows how widening the mirror transistors causes a larger saturation current.*

When a differential input signal of v_{in} is applied the current through M_{Ib} is given by

$$\begin{aligned} I_I &= \beta_1(V_{GSTI,Q}+v_{in})^2 \\ &= I_B(1+x)^2 \qquad (4.39) \\ \text{with} \quad x &= \frac{v_{in}}{V_{GSTI,Q}} \qquad (4.40) \end{aligned}$$

Then, in the left output branch, the single ended current being sourced ($I_{SE,t}$) and the one being sunk ($I_{SE,b}$) (the latter only valid for small applied input voltage) are

$$\begin{aligned} I_{SE,t} &= B(I_B+I_B(1+x)^2) \\ &= BI_B(1+(1+x)^2) \qquad (4.41) \\ I_{SE,b} &= BI_B(1+(1-x)^2) \qquad (4.42) \end{aligned}$$

The latter current is being generated in the other input stage, not shown in the figure. The single ended output current for small input signals is then given by

$$\begin{aligned} I_{SE} &= I_{SE,t}-I_{SE,b} \\ &= 4xBI_B \qquad (4.43) \\ \text{or} \quad y &= 4Bx \qquad (4.44) \\ \text{with} \quad y &= \frac{I_{SE}}{I_{BI}} \qquad (4.45) \end{aligned}$$

So for small input signal excursions, the output current is linear. The single ended output current can also be expressed as

$$I_{SE} = 2Bg_{mI}v_{in} \qquad (4.46)$$

This does not come as a surprise. For a small input signal, the single ended output current is expected to be proportional to the transconductance of a transistor in the input stage, the mirror factor B. The factor of two is caused by the push-pull operation of the amplifier.

For large x (4.42) is not valid. However, the corresponding transistor is turned off, so that term vanishes. Therefore the current is quadratic and given by

$$y = B(1+x)^2 \qquad (4.47)$$

The crossing point between the linear and quadratic region can be found by equating (4.44) and (4.47):

$$\begin{aligned} x_{l\to q} &= 1 \qquad (4.48) \\ \text{or} \quad v_{in,l\to q} &= V_{GSTI,Q} \qquad (4.49) \end{aligned}$$

For very large x the output current saturates. This happens at

$$\begin{aligned} x_{q\to s} &= \left(\sqrt{\frac{y_{max}}{B}}-1\right) \qquad (4.50) \\ \text{or} \quad v_{in,q\to s} &= V_{GSTQ,I}\left(\sqrt{\frac{I_{MAX}}{BI_B}}-1\right) \qquad (4.51) \end{aligned}$$

The saturation condition occurs in practice when the output transistors gate to source voltage reaches $V_{DD}-V_{SS}$. The maximum single ended output current is then

$$I_{SE,MAX} = \frac{KP}{2}\frac{W}{L}(V_{DD}-V_{SS}-V_T)^2 \qquad (4.52)$$

The indices are left out because there are only two output transistors conducting at that moment, and it could be either the *n*MOST or the *p*MOST which saturates first. It is good design practice to size them such that they simultaneously reach their maximum current. In that case the single ended DC transfer characteristic is quite point symmetric.

4.4.2.2 Higher Order Model.

When comparing the above model with simulation results, some discrepancies are found. This more elaborate model takes into account the fact that the input stage is biased slightly in weak inversion and also the threshold voltage variation due to the bulk effect.

The input transistors are biased in the moderate inversion region, i.e., slightly in weak inversion. There are two good reasons to do this:

- In order to bias a transistor in strong inversion a minimum V_{GST} of $200\,mV$ is required. This is impossible if the intended power supply is too low, i.e., $V_{DD} - V_{DSsat,2} < V_T + 200\,mV$.
- The transfer characteristic has a more exponential behavior. Larger output currents occur for a certain input voltage because of the exponential function, and once in the quadratic region, the generated currents are larger due to the larger aspect ratio. Another benefit is the larger g_m/I ratio.

The transfer characteristic can be expressed as

$$I_{SE} = B\, I_{D_0,I}\left(e^{\frac{V_{GSQ,I}+v_{in}}{nU_t}} - e^{\frac{V_{GSQ,I}-v_{in}}{nU_t}}\right) \tag{4.53}$$

$$\text{or} \quad y = B(e^{x} - e^{-x}) \tag{4.54}$$

$$\text{with} \quad y = \frac{I_{SE}}{I_B},$$

$$I_B = I_{D_0,I}\, e^{\frac{V_{GSQ,I}}{nU_t}},$$

$$x = \frac{v_{in}}{nU_t}$$

Linearization of this expression for small x yields

$$y = 2Bx \tag{4.55}$$

$$I_{SE} = 2B g_{mI} v_{in}, \tag{4.56}$$

$$\text{with} \quad g_{mI} = \frac{I_B}{nU_t} \tag{4.57}$$

Equations (4.56) and (4.46) are equal, which is expected.

With growing differential input signal, the V_{GS} of of one input transistor increases while the other decreases. The former transistor at a certain point reaches strong inversion and draws a large current, while the current through the latter can at that point be neglected. Therefore:

$$I_{SE} = \frac{\beta_1}{2}(V_{GSI} - V_{TI})^2 \tag{4.58}$$

$$\text{with} \quad \beta_1 = KP_p\frac{W_I}{L_I}$$

It requires some effort to determine V_{GSTI} in the above expression.

$$I_{SE} = \frac{\beta_1}{2}\Bigg(\underbrace{\left(V_{GQ,I}+\frac{v_{in}}{2}\right)}_{V_G}-\underbrace{\left(V_{GQ,I}-V_{GSTQ,I}-V_{TQ,I}-\frac{v_{in}}{2}+\Delta V_T(\gamma)\right)}_{V_S} - \underbrace{(V_{TQ,I}-\Delta V_T(\gamma))}_{V_T}\Bigg)^2 \quad (4.59)$$

$$= \frac{\beta_1}{2}(V_{GSTQ,I}+v_{in})^2 \quad (4.60)$$

Especially the middle term in this expression may require explanation. Consider the input stage in Fig. 4.5. The source node of M_{Ib} i.e., n_{Ia} is determined by M_{Ia}. It follows its source by a constant V_{GST} plus a V_T which is dependent on the applied input voltage. V_{GST} is constant, because the feedback loop keeps the current through M_{Ia} constant. Further, when an input voltage is applied, V_{SBI} changes which alters V_{TIa} through the bulk factor γ by an amount ΔV_T. Transistor M_{Ib} has the same ΔV_T. For a practical model the bulk effect can be approximated by a linear effect if V_{SB} is not too large, which yields for V_T:

$$V_T \approx V_{T0} + \gamma\frac{V_{SB}}{2\sqrt{|2\Phi_F|}} \quad (4.61)$$

Due to V_T changing the excursion on n_{Ia} is therefore not $v_{in}/2$ but $\xi v_{in}/2$. This can be determined by the following equations for M_{Ib}, for which the V_{GST} must remain constant.

$$\begin{cases} V_{GSQ,I}-V_{TQ,I}=V_{GSIb}-V_{TIb}=C' \\ V_{GSIb}=V_{GSQ,I}-(1-\xi)\frac{v_{in}}{2} \\ V_{TI}=V_{TQ,I}-\frac{\gamma}{2\sqrt{|2\Phi_F|}}\left(\xi\frac{v_{in}}{2}\right) \end{cases} \quad (4.62)$$

It follows that

$$\xi = \frac{1}{1+\frac{\gamma}{2\sqrt{|2\Phi_F|}}} \quad (4.63)$$

All that is left is to determine $V_{GSTQ,I}$ in (4.60). It is given by

$$V_{GSTQ,I} = nU_t ln(\frac{I_B}{I_{D_0}}) - V_{TQ,I} \quad (4.64)$$

$$\text{where} \quad I_{D_0} = I_{D_00}\frac{W_I}{L_I},$$

$$V_{TQ,I} \approx V_{T0} + \frac{\gamma}{2\sqrt{|2\Phi_F|}}(V_{DD}-V_{GSQ,I})$$

The transition between linear and quadratic current is found solving the quadratic equation obtained from equating (4.56) and(4.60) and is given by

$$V_{in,l\to q} \approx \frac{4g_{mI}}{\beta_1} = \frac{4I_B}{nU_t KP_p\frac{W_I}{L_I}} \quad (4.65)$$

The saturation current is also given by (4.51) and it now occurs at

$$V_{in,q\to s} \approx \sqrt{\frac{\beta_{\{3|4\}}}{\beta_1}}(V_{DD}-V_{SS}-V_{T0,\{3|4\}}) - V_{GSTQ,I} \quad (4.66)$$

4.4.3 Frequency Analysis

In the presented class AB OTA of Fig. 4.11 two parallel signal paths can be discerned. An input signal generates a current through input transistors M_{Ib} and M_{Ic}. The current through M_{Ib} is mirrored through the p-type current mirror to out_1. It is also mirrored to out_2 through the n-type current mirror. Adversely, the single ended output current is the sum of two currents, generated by different transistors in the input stage, and reaching the output via two different paths. The transfer functions of these are now first studied separately.

4.4.3.1 Transfer Function of Separate Signal Paths.

The transfer functions from the input voltage to the drain current of either $M_{3a,b}$ and $M_{6a,b}$ in Fig. 4.11 are

$$A_t(s) = \frac{i_3}{v_{in,d}} \approx g_{mI}\frac{\left(1+s\frac{C_I}{2g_{mI}}\right)}{\left(1+s\frac{2C_3}{g_{m2}}+s^2\frac{C_3C_I}{g_{m2}g_{mI}}\right)} \tag{4.67}$$

$$A_b(s) = \frac{i_6}{v_{in,d}} \approx -g_{mI}\frac{\left(1+s\frac{C_3}{g_{m2}}+s^2\frac{C_3C_I}{2g_{m2}g_{mI}}\right)}{\left(1+s\frac{C_4}{g_{m5}}+s^2\frac{C_4C_2}{g_{m5}g_{m7}}\right)\left(1+s\frac{2C_3}{g_{m2}}+s^2\frac{C_3C_I}{g_{m2}g_{mI}}\right)} \tag{4.68}$$

The indices t and b refer to the signal currents flowing in the *top* brach and the *bottom* branch of the output stage. The nodal capacitances are expressed by C_i (where i refers to node n_{ia} or n_{ib}) and are given by

$$\begin{aligned} C_I &= 2C_{GSI} + 2C_{SBI} + C_{DB2} \\ C_2 &= C_{GS7} + C_{SB7} + C_{DB5} + C_{DBI} \\ C_3 &= C_{GS2} + C_{GS3} + C_{DBI} + C_{DB4} \\ C_4 &= C_{GS5} + C_{GS6} + C_{DB7} + C_{DB8} \end{aligned} \tag{4.69}$$

Equation (4.67) is in fact the frequency behavior of the input stage discussed in Section 4.3. Equation (4.68) combines the frequency behavior of the low voltage current mirror of Section 4.2 and the effect of the frequency dependency of the current injected into its input node. The frequency behavior of that current is the one from the input stage through $M_{Ib,c}$.

These transfer functions look quite complicated at first sight, but the following practical design choices allow to further simplify them:

- The dominant pole is associated with the input stage because it is p-type.
- The pole associated with n_I is made much higher than the GBW of the local feedback loop it is in, ie. $\frac{g_{mI}}{C_I} \gg \frac{g_{m2}}{C_3}$ Therefore the poles become real and one pole cancels with a zero.

These choices allow to simplify the transfer functions as follows:

$$A_t(s) = \frac{i_3}{v_{in,d}} \approx g_{mI}\frac{1}{\left(1+s\frac{2C_3}{g_{m2}}\right)} \tag{4.70}$$

$$A_b(s) = \frac{i_2}{v_{in,d}} \approx -g_{mI}\frac{\left(1+s\frac{C_3}{g_{m2}}\right)}{\left(1+s\frac{C_4}{g_{m5}}+s^2\frac{C_4C_2}{g_{m5}g_{m7}}\right)\left(1+s\frac{2C_3}{g_{m2}}\right)} \tag{4.71}$$

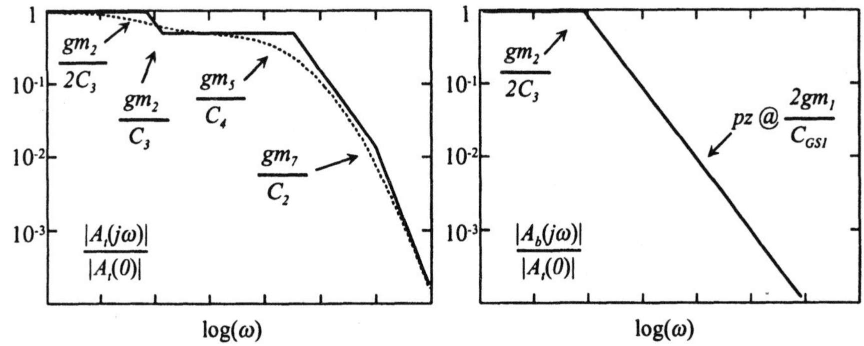

Figure 4.13. *Qualitative Bode plot of the two signal paths of the current output of the class AB OTA. These are only the non-dominant poles and zeros.*

The dominant pole of the top path, generated in the input stage, also appears in the bottom path. In the bottom path, further a zero at the double of the frequency exists, and then the frequency behavior of the low voltage current mirror follows. A sketch of the poles and zeros is shown qualitatively in the Bode plot of Fig. 4.13.

The total transfer function from the differential input voltage to the single ended output current can be found as the difference between the two output currents from the different signal paths:

$$A_{SE}(s) = \frac{i_{out,se}}{v_{in,d}} = A_t(s) - A_b(s) \approx 2g_{m1}\frac{\left(1 + s\frac{1}{2}\left(\frac{C_4}{g_{m5}} + \frac{C_3}{g_{m2}}\right) + s^2\frac{C_4C_2}{2g_{m5}g_{m7}}\right)}{\left(1 + s^2\frac{C_3}{g_{m2}}\right)\left(1 + s\frac{C_4}{g_{m5}} + s^2\frac{C_4C_2}{g_{m5}g_{m7}}\right)} \tag{4.72}$$

The combination of two different signal paths gives rise to the creation of zeros.

Sofar only the non-dominant poles and zeros have been treated. Fig. 4.13, (4.70), (4.71) and (4.72) only take into account non-dominant poles and zeros. The dominant pole is created on the output node of the OTA. The total signal transfer function from differential input voltage to single ended output voltage is given by:

$$H_{SE}(s) = \frac{v_{out,se}}{v_{in,d}} = A_{SE}(s)\frac{1}{g_o + sC_L} \approx \frac{2g_{m1}}{g_o}\frac{1}{\left(1 + s\frac{C_L}{g_o}\right)}\frac{\left(1 + s\frac{1}{2}\left(\frac{C_4}{g_{m5}} + \frac{C_3}{g_{m2}}\right) + s^2\frac{C_4C_2}{2g_{m5}g_{m7}}\right)}{\left(1 + s^2\frac{C_3}{g_{m2}}\right)\left(1 + s\frac{C_4}{g_{m5}} + s^2\frac{C_4C_2}{g_{m5}g_{m7}}\right)} \tag{4.73}$$

The low voltage class AB OTA is to be used in a Switched Capacitor integrator. A requirement for such a system is that it settles to a certain accuracy. Additional poles and zeros influence the settling process. The complicated form of the transfer function of the OTA must not cause worries at all. Complete settling is by all means possible, if certain precautions are taken. This matter is discussed in the next section.

4.4.4 Notes on Settling Behavior

The transfer function of the low voltage class AB OTA contains a considerable amount of poles and zeros as is clear from the previous subsection. Some of these poles and zeros are paired in the sense that they cancel each other out to a certain extent. Others appear as doublets, i.e., the zero occurs at the double of the pole frequency. The OTA is to be used in a Switched Capacitor integrator, which must be able to accomplish complete settling. The occurance of multiple poles and zeros, pole-zero pairs and doublets in a transfer function affect settling, but do not render it impossible if certain precautions are taken.

A treatment of the effect on settling of poles and zeros in a transfer function is given in Appendix C. From it, the following conclusions can be drawn. In the case of pole-zero cancellation at a frequency below the GBW and for real closed loop poles, there is a fast and a slow settling component in the step response. The settling is limited by the time-constant corresponding to the pole-zero frequency. Artificially enlarging the GBW by compensating a non-dominant pole occurring below the GBW with a zero, is not effective. In the case of pole-zero cancellation, which occurs at a frequency higher than the GBW, there are also two settling components. The slowest one is determined by the GBW. The fast one is determined by the pole-zero frequency, and has a coefficient that is proportional to the relative error between the pole and the zero. So, besides being fast, it has also a small coefficient. A negative zero occurring at the double frequency of a pole, as in a doublet, for frequency higher than the GBW, causes a fast settling term with speed determined by the time-constant of the pole. The coefficients of the two terms are equal. Considering the faster decay, it can be concluded that a doublet does not significantly degrade settling. In fact a doublet often allows to double the transconductance of an amplifier, such that the GBW is double and the fast settling term decays with a double time-constant. In the case of complex poles and a zero, a distinction must be made between negative and positive zeros. Negative zeros can be applied to enhance settling. Positive zeros should be avoided, and must be at a frequency much higher than the GBW in order not to degrade settling.

Overall it can be said that the occurance of many poles and zeros in whatever combination in an amplifier must not lead to the misconception that complete settling is impossible,at least if they occur at frequencies higher than the GBW.

4.4.5 Design Strategy

In order to dimension the transistors of the OTA which is a rather complex circuit with many degrees of freedom a systematic procedure is required. The following procedure has been developed for the implementation of the SO integrator of the realization of Section 6.2.

1. Given the supply voltage, determine the quiescent input transistor gate source voltage and its required bias current and/or transconductance from system level analysis, making use of the models derived of the OTA. For the former a good starting point is $V_{GS1,Q} = V_{DD} - V_{in,d,max}$. Later it needs to be checked that all transistors of the input stage remain in saturation under maximum input signal conditions.

2. Determine the ratio of the aspect ratios of the output transistors M_3 and M_6 by requesting equal saturation currents for both transistors, which makes the single ended transfer characteristic point symmetrical. This yields:

$$\frac{\left(\frac{W_3}{L_3}\right)}{\left(\frac{W_6}{L_6}\right)} = \frac{KP_6(V_{DD} - V_{SS} - V_{T6})^2}{KP_3(V_{DD} - V_{SS} - V_{T3})^2} \tag{4.74}$$

   Further choose M_2 equal to M_3 and M_5 equal to M_6. This maximizes the frequency of operation of the input stage and low voltage current mirror.

3. Determine the absolute values of the aspect ratios. First determine the maximum effective input voltage of the OTA. This may need a feed-forward analysis as in Section 5.1.1 if used in an integrator. This maximum input signal determines the maximum current through M_2 and M_5. Under this condition M_4 and M_8 must remain in saturation. Hence the maximum allowed V_{GS} of M_2 and M_5 can be determined

$$\begin{cases} V_{GS3@I_{max}} = V_{DD} - V_{DSsat,4} \\ V_{GS5@I_{max}} = V_{DD} - V_{DSsat,8} \end{cases} \tag{4.75}$$

From the maximum allowed V_{GS} and the aspect ratio already determined, the absolute value of the aspect ratios can be determined

$$\left(\frac{W}{L}\right) = f(I_{max}, V_{GSmax}) \tag{4.76}$$

This dependence can deviate from the MOST square law relationship, in case the supply voltage is so low that the maximum V_{GS} requires operation in the transition region between weak and strong inversion. In that case, a first estimate can be found by using the square law relationship, while a few iterations with a circuit simulator using the full model of the device, can yield a more accurate result.

4. Check that all transistors of the input stage remain in saturation at all times, i.e., when the maximum input signal is applied.

$$V_{DD} - V_{DSsat,2,max} \geq V_{GS1,Q} + \xi V_{in,se,max} \tag{4.77}$$

$$V_{GS1,Q} + (2-\xi) V_{in,se,max} - V_{DSsat,1,Q} \geq V_{DD} - V_{DSsat,2,min} \tag{4.78}$$

The factor ξ is defined in Section 4.4.2.2 on page 76. The former one of these conditions expresses that under the input condition for which the maximum input signal is drawn in the input transistor $M_{1b,c}$, transistor M_2 must stay in saturation. Indeed the saturation voltage of M_2 increases a lot, while the source of the input transistors is pushed upwards. The latter one of these conditions expresses that when the input transistor $M_{1b,c}$ is pinched of, $M_{1a,d}$ remains in saturation. Indeed, then the current through M_2 drops by a factor of two and its gate moves upwards, while the source of $M_{1a,d}$ moves downwards, so removing the condition for saturation. In the case that one of the conditions is not met, a new $V_{GS1,Q}$ needs to be chosen. In the case that none of the conditions are satisfied, the maximum current ratio (ratio of maximum current and quiescent current) has been chosen too high. A smaller maximum current ratio needs to be effected. This is most easily done by choosing a higher quiescent bias current. The procedure must be followed starting from 2.

5. Determine the area of the transistors M_2 and M_3 by setting the non dominant pole. In a Switched Capacitor integrator, which is a feedback system, the *GBW* is determined by the effective transconductance $g_{m,eff}$ and the effective load capacitance $C_{L,eff}$, if the OTA is load compensated as is the case here. *Load compensated* means that the frequency compensation to stabilize the amplifier is provided by the load capacitance. For best settling a critically damped system is desired. This requires a ratio of 4 between the first non dominant pole and the *GBW*. The first non dominant pole is most easily realized as $\frac{g_{m2}}{2C_3}$. By keeping the ratio of $M_{2,3}$ fixed, but increasing the area, the pole can be positioned, as a fixed ratio doesn't alter the transconductance.

6. Cancel the zero at the double of the non-dominant pole frequency with the next pole, which can also be done by keeping the aspect ratio of $M_{5,6}$ fixed, but by altering their area such that:

$$\frac{g_{m5}}{C_4} = \frac{g_{m2}}{C_3} \tag{4.79}$$

As the frequency of this pole-zero pair is at about eight times the GBW the settling behavior is hardly affected.

7. At this point it must be checked that all transistors involved in the bottom current mirrors remain in saturation in all conditions.

$$V_{bias} - V_{GS7} \geq V_{DSsat,5,max} \tag{4.80}$$
$$V_{bias} - V_{GS7} \leq V_{GS1,Q} - (2-\xi)V_{in,se,max} - V_{DSsat,1,max} \tag{4.81}$$
$$V_{GS5,min} \geq V_{bias} - V_{GS7} + V_{DSsat,7} \tag{4.82}$$

The former one of these conditions ensures that M_{5a} is always in saturation, even under maximum input conditions. The second one does the same for $M_{1b,c}$ as in maximum input conditions the source of one of these is moving down, while its drain is staying at a fixed voltage. The third ensures that M_7 is always in saturation.

8. Check that the LF gain is satisfactory. Enlarging the transistors in the output branches M_3 and M_6 and M_2 and M_5, while keeping the same aspect ratio, has a beneficial effect on the gain of the OTA. It comes at the cost of reducing the non-dominant pole in frequency. The gain can be increased more, but the OTA becomes much slower. If the parasitic poles and zeros are maximized in frequency, a very limited gain results.

9. Optimize the size of M_7. The pole associated with its transconductance is maximized in frequency. For maximum frequency its length is chosen minimum feature size. The current through this transistor is constant. Hence its transconductance is proportional to the square root of its width. Varying W_7 changes the nodal capacitances of C_4 slightly through its C_{DB} and C_5 through its C_{GS}:

$$C_2 = C_{2,\alpha} + C_{2,\beta} W_7 \tag{4.83}$$
$$C_4 = C_{4,\alpha} + C_{4,\beta} W_7 \tag{4.84}$$

Consequently two poles are affected. Optimizing W_7 then comes down to minimizing the phase contribution of these two poles at the frequency of the GBW.

$$\Delta\phi_{@GBW} = \arctan\left(\frac{\omega_{GBW}}{\frac{g_{m5}}{C_4}}\right) + \arctan\left(\frac{\omega_{GBW}}{\frac{g_{m7}}{C_2}}\right) \tag{4.85}$$

Using this procedure it is possible to obtain integrators with very decent Bode plots and good settling behavior. A design example is given in Section 6.2.

4.4.6 Noise

The equivalent input referred noise of the OTA can be expressed as

$$\begin{aligned} P_{N,eq} \; = \; & \frac{2kT}{3C_{L,eff}}(\Gamma_{w,OTA} + \Gamma_{w,cmfb}) \\ & + \frac{KF\, I_{DS1}{}^{AF}}{C_{ox} g_{m,eff}^2 L_{1,eff}^2} ln\left(\frac{f_c}{f_l}\right)(\Gamma_{f,OTA} + \Gamma_{f,cmfb}) \end{aligned} \tag{4.86}$$

In this the first term is the white noise and the second term is the $1/f$ noise. The factors $\Gamma_{w,OTA} + \Gamma_{w,cmfb}$ and $\Gamma_{f,OTA} + \Gamma_{f,cmfb}$ are the excess noise factors and indicate how much more noise the amplifier has than that generated by the transistors contributing to the effective transconductance, i.e., the input transistors. The terms $\Gamma_{w,cmfb}$ and $\Gamma_{f,cmfb}$ are due to differential mode noise generated in the CMFB, and are calculated in Section 4.5.5. In order to determine the excess noise factors, the contribution of each transistor is determined. The equivalent input noise voltage density is calculated as

$$\overline{dv_{nie}^2} = \sum_{k=1}^{n} \overline{di_{nds,k}^2} \left| \frac{H_{ko}(s)}{H_{io}(s)} \right|^2 \tag{4.87}$$

It is the sum of the contribution of the drain source noise currents of each transistor to the output, each divided by the transfer function from the input to the output. As the amplifier is load compensated, the frequency dependence can be left out. The equivalent input noise density can be rewritten as

$$\overline{dv_{nie}^2} = \frac{\sum_{k=1}^{n} \overline{di_{nds,k}^2} \, |H_{ko}|^2}{|H_{io}|^2} \tag{4.88}$$

$$= \frac{1}{g_{meff}^2} \sum_{k=1}^{n} \overline{di_{nds,k}^2} \, |H_{ko}|^2 \tag{4.89}$$

$$= \frac{1}{g_{meff}^2} \sum_{k=1}^{n} \overline{di_{no,k}^2} \tag{4.90}$$

$$= \frac{1}{g_{meff}^2} (\overline{di_{no,w}^2} + \overline{di_{no,f}^2}) \tag{4.91}$$

Because the OTA makes ample use of local current feedback schemes, it is more natural to consider drain source noise currents. Table 4.3 shows the noise contributions of each of the transistors at the output, separately for the white noise ($\overline{di_{no,w}^2}$) and the $1/f$ noise ($\overline{di_{no,f}^2}$). A qualitative understanding of noise contributions is given for the white noise sources. (The $1/f$ noise is similar). The noise current of M_{1b} is injected in parallel with the input signal current. It appears differentially and therefore causes correlated differential noise at the output. This explains the factor of 4. Transistor M_{1a} is the only one more easily studied considering its equivalent input noise voltage. As its source voltage determines the current through M_{1b}, its input noise voltage causes a noise current in M_{1b}, which also appears differentially. An analytical derivation of this is provided in Appendix B.2. The noise current of M_4 flows through M_{1a} into M_2 and is mirrored to the output. However it changes the v_{GS} of M_{1a}, which causes a noise current of equal magnitude in M_{1b}, which appears at both outputs and is added coherently. This explains the factor of 9. The noise current of M_2 is injected into the low impedance node of the current buffer structure of M_{1a}, M_2 and M_3 and appears at one output. The noise current of M_3 also adds to one output. The noise currents of M_5 and M_6 are analogous. Also the noise current of M_8 is mirrored to one output. This leaves M_7. It can be regarded as a cascode transistor that sees a low impedance at its drain, in which case the noise is suppressed and negligible. Nevertheless a derivation is done in Appendix B.1.

A few remarks can be made concerning the $1/f$ noise. Normally $KF_p << KF_n$, and therefore only the nMOST need to be considered. Further, from the design procedure of Section 4.4.5 it follows in practice that all nMOST transistor lengths are much larger than minimal size except L_4. Hence, in a practical situation M_4 is the only transistor really contributing to the $1/f$ noise.

It is interesting to obtain an idea of Γ_w. As a first approach, all transconductances can be assumed to be proportional to the bias current only (weak inversion biasing or constant V_{GS}).

	$\overline{di^2_{o,w}}$	$\overline{di^2_{o,f,1Hz}}$
	$\times \frac{8}{3}kT$	$\times \frac{I^{AF}_{B,l}}{C_{ox}}$
M_{1a}	$4 \times g_{ml}$	$4 \times \frac{KF_p}{L1,eff}$
M_{1b}	$4 \times g_{ml}$	$4 \times \frac{KF_p}{L1,eff}$
M_2	g_{m2}	$2^{AF} \frac{KF_p}{L2,eff}$
M_3	g_{m3}	$2^{AF} \frac{KF_p}{L3,eff}$
M_4	$9 \times g_{m4}$	$4 \times \frac{KF_p}{L4,eff}$
M_5	g_{m5}	$2^{AF} \frac{KF_n}{L5,eff}$
M_6	g_{m6}	$2^{AF} \frac{KF_n}{L6,eff}$
M_7	$\left(\frac{g_{m5}+[1]}{g_{m7}}\right)^2 g_{m7}$	$\left(\frac{g_{m5}+[1]}{g_{m7}}\right)^2 \frac{KF_n}{L7,eff}$
M_8	g_{m8}	$\frac{KF_n}{L8,eff}$

Table 4.3. *White and $1/f$ noise contributions of each transistor of the class AB OTA.*

This is an unrealistic assumption, but it gives an order of magnitude. To find Γ_w all entries in the first column of Table 4.3 are summed. The transconductances can be expressed as a multiple of g_{ml}, depending on the ratio of the bias currents of that particular transistor to M_l. Dividing by the effective differential transconductance $4g_{ml}$ yields Γ_w. Doing so, one finds a Γ_w of 13. In a real implementation (see Section 6.2), it turns out to be

$$\Gamma_w \approx 10 \tag{4.92}$$

This is a relatively high number.

4.5 CMFB Circuitry

The error amplifier based approach of Section 3.4.4.2 was chosen for implementation. Error amplifier based CMFB systems have proven to be very robust. It also provides an extra flexibility in the design due to the fact that it allows its DC bias conditions to be decoupled from the DC bias of the integrating opamp, by feeding in AC feedback current in parallel with the input signal. It also has the very important advantage of the immediate charge redistribution to the correct common mode voltage, as explained in Section 3.4.4.5.

4.5.1 The Trifferential Input Stage

The output common mode voltage of the low voltage class AB OTA of Fig. 4.11 can be regulated through several control points. The first option is the current sources M_4. However, these bias the input stage, and doing the common mode control through these current sources, would vary

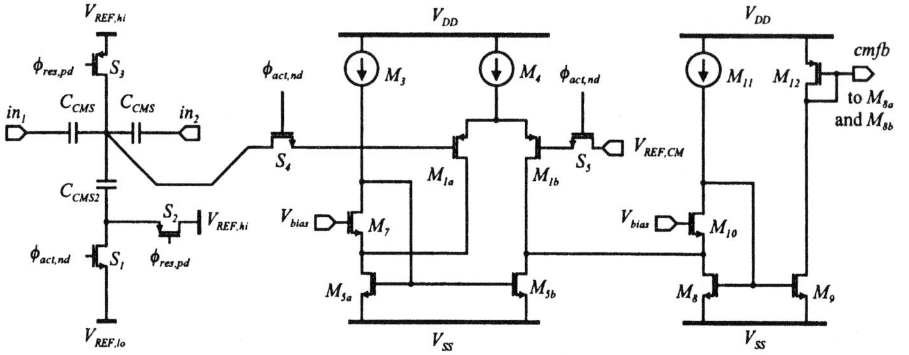

Figure 4.14. *An error amplifier based CMFB with feedback to a current mirror. An intermediate block is needed to properly apply the feedback signal, but increases the number of nodes in the feedback signal path.*

the properties of the input stage, such as the transconductance. The consequence is unreliable behavior due to possibility of instability and is therefore ruled out.

The second option is to apply the common mode feedback system through the current sources M_8. Fig. 4.14 shows the needed circuitry. On the left of the figure the common mode sampling network introduced in Section 3.4.4.2 appears. The next block is the error amplifier itself. It consists of a p-type input differential pair, with a single ended output. The left branch makes use of the low voltage current mirror discussed in Section 4.1.2. The single ended output error current cannot be injected into the output node directly, because it doesn't have rail-to-rail output swing, and even then, it would lower significantly the output conductance of the integrator OTA, and limit the maximum obtainable OTA gain. Since the OTA is a one stage amplifier it is important to avoid this. In order to apply the error current to the output node, the detour via M_8 is taken. The block on the right of the figure shows the necessary circuitry to do this. The disadvantage of this approach is immediately apparent. From the output of the error amplifier to the output node of the class AB OTA, there are three nodes. They are n_{14} and n_{15} of the intermediate block and n_2 of the class AB OTA. The poles associated make it difficult to achieve high GBW for the CMFB loop.

There is, however, a third option to carry out the feedback. Provided some modifications, it is possible to inject the error current into node n_2 of the OTA, which is therefore also called '*cmfb*'. Doing so the detour via M_8 and the intermediate block at the right of Fig. 4.14 can be avoided. If the output node of the error amplifier would be directly connected to the nodes n_{2a} and n_{2b}, obviously those nodes would be shortened, which would destroy the class AB behavior of the OTA. These two nodes must be isolated from each other. This can be done by splitting the transistors in the output branch in two, as shown in Fig. 4.15. This operation gives rise to a input structure with three input transistors instead of two. Hence it is a sort of "triple differential" stage, and is therefore here given the name "trifferential input stage". Every CMFB system has a point where the feedback path is split to be directed to one output. Here, this split is in the error amplifier itself, rather than in the OTA.

Attention must be paid to switches S_4 and S_5 which are connected to the gates of the input transistors $M_{1a,b,c}$. These switches are not strictly necessary. They help to reduce the systematic common mode offset error caused by charge flowing to the parasitic capacitance represented by the C_{GS} of the input transistors. The ones on the side of the reference voltage are not even

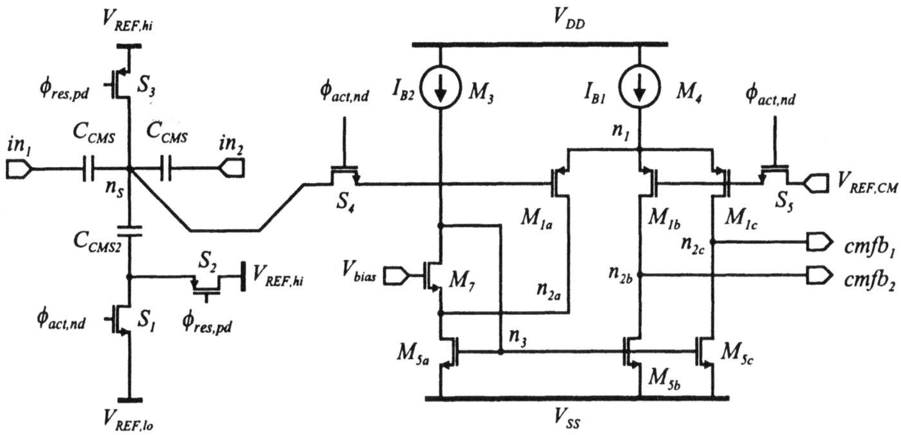

Figure 4.15. *An error amplifier based CMFB with feedback to nodes n_{2a} and n_{2b} (in Fig. 4.11) in parallel with differential signal. This requires the output branch of the error amplifier to be split in to in order to isolate the mentioned nodes from each other.*

needed for that reason. Keeping the C_{GS} charge trapped on both sides of the input stage, keeps the latter balanced during the off stage.Hence, when going active again, the input stage doesn't systematically steer current (potentially in the wrong direction), due to the different $V_{GS1a,b,c}$.

4.5.2 Frequency Analysis

Here the transfer function of the trifferential stage is derived for an input applied at the gate of M_{1a}. It is done in three steps. First the transfer function from the input signal to the drain currents of each of the transistors is derived. Then, given the transfer function of the current mirror, the overall transfer function can be calculated. In this approach M_{1b} and M_{1c} are assumed to be identical with identical bias current. This is a special case, but it is required in this application. Therefore g_{m1b} equals g_{m1c}.

The transfer functions from a small signal input voltage applied to the drain currents i_{1a}, i_{1b}, i_{1c} are calculated in Appendix B.3 and are given by:

$$H_a(s) = \frac{i_{1a}}{v_{in}} = \left(\frac{g_{m1a}2g_{m1b}}{g_{m1a}+2g_{m1b}}\right)\left(\frac{1+s\left(\dfrac{2C_{GS1b}+C_p}{2g_{m1b}}\right)}{1+s\left(\dfrac{C_{GS1a}+2C_{GS1b}+C_p}{g_{m1a}+2g_{m1b}}\right)}\right) \quad (4.93)$$

$$H_b(s) = \frac{i_{1b}}{v_{in}} = -\left(\frac{g_{m1a}g_{m1b}}{g_{m1a}+2g_{m1b}}\right)\left(\frac{1+s\left(\dfrac{2C_{GS1a}}{2g_{m1a}}\right)}{1+s\left(\dfrac{C_{GS1a}+2C_{GS1b}+C_p}{g_{m1a}+2g_{m1b}}\right)}\right) \quad (4.94)$$

Capacitor C_p is the parasitic capacitance seen at the source of the input transistors.

$$C_p = \sum_{k=a}^{c} C_{SB1k} + C_{DB4} + C_{GD4} \tag{4.95}$$

The transfer function for the low voltage current mirror is found from (4.14)

$$H_{(cm)}(s) = \frac{i_{out}}{i_{in}} \approx -\frac{g_{m5b}}{g_{m5a}} \frac{1}{(1 + s\frac{C_3}{g_{m5a}} + s^2\frac{C_{2a}C_3}{g_{m7}g_{m5a}})} \tag{4.96}$$

The transfer function from input voltage to output current is then

$$\begin{aligned} H_{tr} &= H_a(s)H_{cm}(s) + H_b(s) \qquad (4.97) \\ &= -\frac{g_{m1a}g_{m1b}}{g_{m1a} + 2g_{m1b}} \cdot \\ &\frac{1}{\left(1 + s\left(\frac{C_{GS1a} + 2C_{GS1b} + C_p}{g_{m1a} + 2g_{m1b}}\right)\right)\left(1 + s\frac{C_3}{g_{m5a}} + s^2\frac{C_{2a}C_3}{g_{m7}g_{m5a}}\right)} \\ &\left\{\left(2\frac{g_{m5b}}{g_{m5a}} + 1\right) + s\left(2\frac{g_{m5b}}{g_{m5a}}\frac{2C_{GS1b} + C_p}{2g_{m2}} + \frac{C_{GSa}}{g_{m1}} + \frac{C_3}{g_{m5a}}\right) + \right. \\ &\left. s^2\left(\frac{C_{GSa}C_3}{g_{m1}g_{m5a}} + \frac{C_{2a}C_3}{g_{m7}g_{m5a}}\right) + s^3\left(\frac{C_{GSa}C_{2a}C_3}{g_{m1}g_{m7}g_{m5a}}\right)\right\} \qquad (4.98) \end{aligned}$$

The index *tr* stands for *trifferential.* In practice this transfer function can be simplified to a one pole system because the *GBW* of the low voltage current mirror turns out to be very very dominant.

$$H_{tr} \approx -\frac{g_{m1a}g_{m1b}}{g_{m1a} + 2g_{m1b}}\left(2\frac{g_{m5b}}{g_{m5a}} + 1\right)\frac{\left(1 + s\frac{C_3}{g_{m5a}\left(2\frac{g_{m5b}}{g_{m5a}} + 1\right)}\right)}{\left(1 + s\frac{C_3}{g_{m5a}}\right)} \tag{4.99}$$

The constant term is the effective transconductance from the input to a single output. The dominant pole creates a zero since it occurs in only one signal path. When the two signal paths are equal one expects to find the zero at the double of the pole frequency. [L&S d] Here the two signal paths are not equal, but differ by a factor of $\frac{g_{m5b}}{g_{m5a}}$ Therefore the zero occurs at a higher frequency. The use of a current mirror is that the full transconductance of the input pair can be regained. This comes at the cost of a pole and a zero. However, by choosing a large mirror factor m, these can be split. This means in fact that the dominant pole is shifting to lower frequencies, but as long as this is allowable, it is good practice.

There are several different biasing schemes possible for the error amplifier. Two are shown in Fig. 4.16. The difference is in the mirror factor m from M_{5a} to M_{5b} and M_{5c}. The transconductance of the input transistors is proportional to the bias current through them:

$$g_{m1a} = \rho I \tag{4.100}$$

$$g_{m1b} = g_{m1c} = \rho m I \tag{4.101}$$

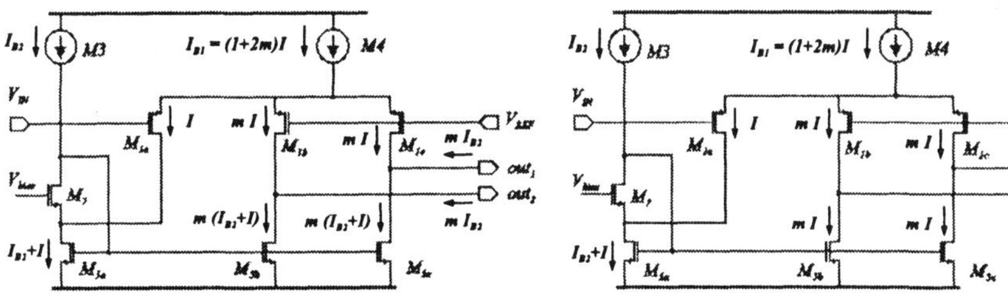

(a) (b)

Figure 4.16. *There are several biasing schemes possible for the CMFB error amplifier. They differ in the mirror factor applied to M_{5a} and $M_{5b,c}$.*

Hence it is possible to calculate the effective transconductance as follows

$$g_{meff,a} = \rho m I \tag{4.102}$$

$$g_{meff,b} = \rho m I \left(\frac{1 + \dfrac{I_{B2}}{(2m+1)I}}{1 + \dfrac{I_{B2}}{I}} \right), \tag{4.103}$$

where (a) and (b) correspond to the two cases of Fig. 4.16. The case (a) is given for a rather didactic reason, because it shows that the overall transconductance is in principle independent of the scaling of the branch with M_{1a}. This would be also the case for a single transistor current mirror (Fig. 4.2), where I_{B2} would be by definition equal to zero. This means that that branch can be scaled down. Scaling it down lowers the frequency of the pole associated with. Hence it starts limiting the frequency responce if scaled down to much. Case (a) is not so interesting, because causes higher current consumption, due to the fact that also I_{B2} is mirrored by the scaling factor m. The total bias currents for both cases are

$$I_{t,a} = (2m+1)(I + I_{B2}) \tag{4.104}$$

$$I_{t,b} = (2m+1)I + I_{B2} \tag{4.105}$$

Case (b) is of more practical interest, because the total current consumption can be less for a given effective transconductance.

Finally note that there is a net DC current flowing into the output node that must be delivered by $M_{8a,b}$ through $M_{7a,b}$ of the OTA in the case of (a). No DC current is flowing in the biasing scheme of (b). The DC biasings of the OTA and the CMFB are decoupled. The CMFB can be designed with more flexibility than if they were coupled. In particular the transconductance, for an important part determining the settling speed of the common mode, is a function of the biasing currents in the error amplifier, and can be set independently of the OTA biasing.

4.5.3 Settling Speed

The CMFB system including all loading capacitors is shown in Fig. 4.17. The load capacitances seen are those of all the capacitors to ground. Those are of course of the capacitors at the output of the differential amplifier C_L and $C_{INT,BP}$ which is the bottom plate capacitor of the integrating

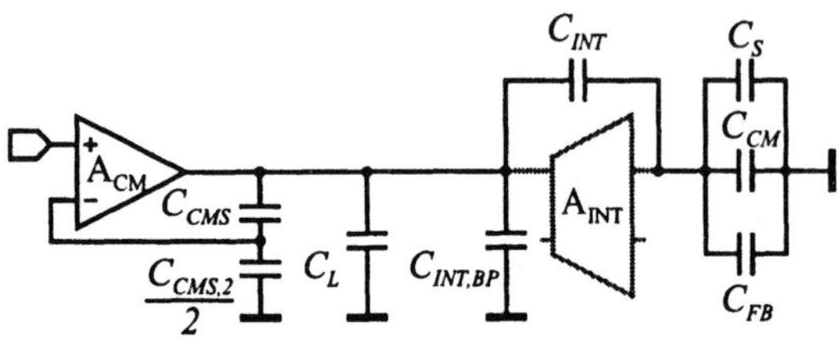

Figure 4.17. *The capacitive loading of CMFB in closed loop. The differential opamp is inactive in common mode, but the capacitors around it are seen by the CMFB. The feedback factor is determined by the common mode sample division factor α.*

capacitor. Besides these, also a load to the CMFB are all the capacitors at the differential amplifier's input (C_S +C_{CM} +C_{FB}) in series with the integration capacitor C_{INT}. Indeed the differential amplifier does not play a role in common mode. The settling speed is determined by the AC behavior. The CMFB amplifier itself is made up of the error amplifier of Fig. 4.15 and shares the bottom output section of the OTA of Fig. 4.11. More specifically the poles related with nodes $n_{2a,b}$ and $n_{4a,b}$ and the output node are in the CMFB signal path. The structure is, however, load compensated. The settling speed is determined by the closed loop GBW, given by

$$
\begin{aligned}
GBW &= \frac{g_{meff,CM}}{2\pi C_{L,eff,CM}} \\
C_{L,eff,CM} &= \frac{C_{CMS2}}{2} + \frac{C_L + C_{BP} + \dfrac{(C_S + C_{FB} + C_{CM})C_{INT}}{(C_S + C_{FB} + C_{CM} + C_{INT})}}{\mathcal{F}_{dc,CM}} \\
\mathcal{F}_{dc,CM} &= \frac{C_{CMS}}{C_{CMS} + \frac{C_{CMS2}}{2}} = \alpha
\end{aligned}
\tag{4.106}
$$

where

C_{CMS} and C_{CMS2}	are the common mode sampling capacitors
C_L	is the integrator's load capacitor
C_{INT}	is the integrating capacitor
C_{BP}	is the bottom plate capacitor of C_{INT}
C_S	is the integrator's sampling capacitor
C_{CM}	is the common mode level shift capacitor
C_{FB}	is the ΔΣ feedback sampling capacitor

The effective transconductance is calculated in the previous paragraph. The expression for the load capacitance is remarkable. It has many terms. In fact the CMFB amplifier sees as a load all capacitors connected to the output of the OTA. The integrator's OTA is differential and cannot influence the common mode behavior. The series capacitance of the integrating capacitor C_{INT} across the OTA and all the capacitors connected to its input node (C_S,C_{FB},C_{CM}) is also seen as a load. Furthermore, the effective load capacitance is also boosted by a feedback factor that so happens to be exactly equal to the common mode division factor introduced in Section 3.4.4.2, which is always smaller than unity. Therefore the effective load capacitance is increased by the feedback factor which is actually the division factor.

It can be concluded that the CMFB amplifier sees quite a large capacitive load that is even effectively increased by the common mode division factor.

4.5.4 The Common Mode Sampling Division Factor

The requirements for C_{CMS} and C_{CMS2} have been defined earlier:

- The equivalent common mode sampling capacitance $C_{CMS,eq}$ seen from nodes n_{in1} and n_{in2} must be equal to the total load capacitance $C_{L,t}$ from the OTA integrator output to ground. (Section 3.4.4.4)
- The voltage sampled on node n_s must satisfy

$$V_s = \alpha \frac{V_{in1} + V_{in2}}{2} \tag{4.107}$$

These conditions are satisfied if

$$\frac{C_{CMS2}}{C_{CMS}} = 2\left(\frac{1}{\alpha} - 1\right) \tag{4.108}$$

$$C_{CMS} = \frac{1}{1-\alpha} C_{CMS,eq} \tag{4.109}$$

$$C_{CMS2} = \frac{2}{\alpha} C_{CMS,eq} \tag{4.110}$$

The above equations allow to determine the relative and absolute values for the capacitacnces to be used in the common mode sampling network.

4.5.5 Noise

It may sound unusual to be concerned about noise of the CMFB system. Noise of the CMFB translates into a noisy common mode voltage, which is irrelevant as long as the differential signal processing is not affected. However, because the presented CMFB error amplifier has two separate outputs feeding into two different signal paths of the OTA, there is also differential mode noise generated by it. Here that noise is calculated. As for the noise of the OTA the drain-source noise currents of the transistors are considered.

When looking at the error amplifier in Fig. 4.15 some transistors that do not cause differential mode noise can immediately be identified and must not be considered further. They are M_{1a}, M_3, M_4, M_{5a} and M_7. Now consider the drain source noise current i_{n1b} of M_{1b}. It is injected into node n_1. There it sees the three low source impedances g_{m1a}, g_{m1b} and g_{m1c} of the trifferential stage transistors. The currents divide between the three branches as follows:

$$i_{1a} = \frac{g_{m1a}}{g_{m1a} + g_{m1b} + g_{m1c}} i_{n1b} = \frac{1}{2m+1} i_{n1b} \tag{4.111}$$

$$i_{1b} = i_{1c} = \frac{g_{m1b}}{g_{m1a} + g_{m1b} + g_{m1c}} i_{n1b} = \frac{m}{2m+1} i_{n1b} \tag{4.112}$$

where m is defined as the ratio of the currents between M_{1a} and $M_{1b,c}$. It also causes a current in $M_{5b,c}$ of

$$i_{5b,c} = i_{1a} q \tag{4.113}$$

where q is the mirror factor between M_{5a} and $M_{5b,c}$. This single noise current generates a differential output (at the output of the OTA) of

$$i_{n1b,o} = (i_{n1b} - i_{1b} + i_{5b}) - (-i_{1c} + i_{5c}) = i_{n1b} \tag{4.114}$$

	$\overline{di^2_{o,w}}$	$\overline{di^2_{o,f,1Hz}}$
	$\times\frac{8}{3}kT$	$\times\frac{I^{AF}_{B,1b,c}}{C_{ox}}$
M_{1b}	g_{m1b}	$\frac{KF_p}{L1b,eff}$
M_{1c}	g_{m1c}	$\frac{KF_p}{L1c,eff}$
M_{5b}	g_{m5b}	$\frac{KF_n}{L5b,eff}$
M_{5c}	g_{m5c}	$\frac{KF_n}{L5c,eff}$

Table 4.4. *White and $1/f$ noise contributions of each transistor of the CMFB contributing to differential OTA noise.*

The noise current of M_{1b} and by analogy M_{1c} appears fully as differential noise current. Finally, the noise currents of $M_{5b,c}$ also appear as differential output noise current.

$$\overline{di^2}_{ncm,w,o} = \frac{8}{3}kT\left(g_{m1b} + g_{m1c} + g_{m5b} + g_{m5c}\right) \tag{4.115}$$

A noise excess factor $\Gamma_{w,cm}$ can be defined, quantifying the excess differential noise due to the CMFB.

$$\Gamma_{w,cm} = \frac{\left(g_{m1b} + g_{m1c} + g_{m5b} + g_{m5c}\right)}{g_{meff,OTA}} \tag{4.116}$$

For the realization of Section 6.2 $\Gamma_{w,cm}$ is approximately equal to three. The $1/f$ noise can be treated analogously. A summary is presented in Table 4.4

4.6 Switching Strategies

In this section switching methods for the opamps are discussed. The original switching method and a new one, which is faster, are treated.

4.6.1 Original Method

The philosophy in the switching method of the Original Switched Opamp Technique is to add as little extra transistors as possible. The resulting switching scheme for the two stage Miller OTA is shown in Fig. 4.18 [Cro 94]. It is switched with its bias current. The switch transistor M_{10} short-circuits n_6 to V_{DD} and thus pulls the V_{GS} of the current mirror made up of M_8, M_7 and M_5 to zero. So a low clock signal ϕ removes the bias currents to the OTA and the OTA goes off. A low clock signal, opens switch M_{10}, such that the current mirror transistor M_8 conducts again. Switch M_9 is added in order not to discharge the compensation capacitor C_c during the switching transitions or the off phase. This precaution must be taken, because the turn on time would be largely increased if C_c is discharged. A very low operation speed of the Switched Opamp circuit would result. When turning off the bias current through M_5, amplifying transistor M_6 does not immediately stop conducting, if it does at all. Its conduction regime depends on V_{GS6}. Capacitor C_{GS6} can be partially discharged by M_4, before the latter is turned off by the discharge of M_3. It cannot be guaranteed that C_{GS6} completely discharges and

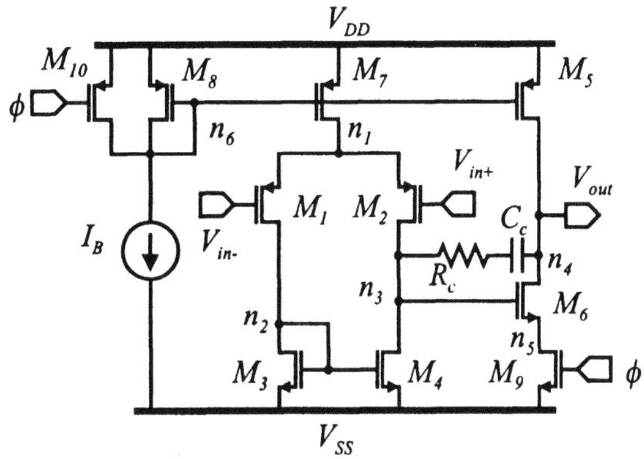

Figure 4.18. *The original switching method in the Switched Opamp Technique consists in switching on and off the bias currents and taking measures such that large capacitances are not discharged.*

it will consequently continue conducting during the off transition, discharging C_c. Therefore switch M_9 is added. Its size is determined by two requirements. Firstly, the voltage drop over it must be small in the on state, because it reduces the output signal swing by twice its value. Secondly, the switch resistance may not be allowed to degenerate the amplification of the output stage significantly. Therefore, g_{ds9} must be much larger than g_{m6}.

Turning on the OTA requires transistor M_8 to be turned on again, which means that V_{GS8} must be established. For this, I_B must charge the load capacitance of M_8 which is its own C_{GS} and also the C_{GS} of the transistors it biases. Below V_{T8}, V_{GS8} slews and then it settles with a time-constant, depending on a transconductance value which varies. The turn-on speed of the current mirror can be approximated by

$$T_{on8} = \frac{V_{T8}(C_{GS8} + C_{GS7} + C_{GS5})}{I_B} + 3\tau_8 \quad (4.117)$$

$$\tau_8 = \frac{(C_{GS8} + C_{GS7} + C_{GS5})}{g_{m8}} \quad (4.118)$$

$$g_{m8} \approx \frac{2I_B}{V_{GST8,Q}} \quad (4.119)$$

When the bias current flows, the OTA is turning on, which requires some more time. Indeed, some internal capacitors that were discharged during the off state, must be recharged. There are several cases to consider. The first are current mirrors. In the off state, the C_{GS} of all current mirror transistors discharge through the diode connected transistor. In Fig. 4.18, C_{GS3} and C_{GS4} discharge through M_3. These capacitors must be recharged at turn on. Transistor M_6 potentially fully discharges its C_{GS} through C_c, regardless of a partial discharge through M_4. The degree of discharge depends much on the voltage across C_c at the end of the specific phase. C_{GS6} must be recharged as well. The voltage across the gate source terminals of the input transistor M_1 depends on the external integrating capacitor, which is connected in series with C_{GS1}. Depending on the voltage stored on C_{INT}, M_1 may be on or off. M_2, however, is always on. If they are both on, their C_{GS} remain charged. If M_1 goes off, its charge redistributes to C_{GS2}. Upon turn on, the inverse redistribution time can be neglected. Therefore, C_{GS1} and C_{GS2} do not have to be considered.

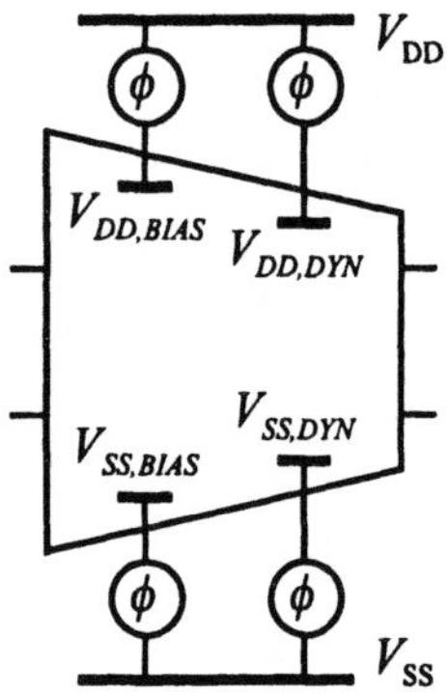

Figure 4.19. *The fast method of switching amplifiers consists in interrupting the current path from the supplies by a switch.*

In general there are several paths to turn on. Some transistors are turned on sequentially, others in parallel. A worst case estimate of the overall turn-on time is given by

$$T_{on,OTA} \approx T_{on8} + max\{T_{on3}, T_{on6}\} \quad (4.120)$$

$$T_{on3} \approx \frac{V_{T3}(C_{GS3} + C_{GS4})}{I_{B,7}/2} + 3\tau_3 \quad (4.121)$$

$$\tau_3 \approx \frac{(C_{GS3} + C_{GS4})}{g_{m3,Q}} \quad (4.122)$$

$$T_{on6} \approx \frac{V_{T6}C_{GS6}}{I_{B,7}/2} + 3\tau_6 \quad (4.123)$$

$$\tau_6 \approx \frac{C_{GS6}}{(g_{o2} + g_{o4})} \quad (4.124)$$

The limiting factor is M_6, due to the large time constant by which it needs to be charged, if fully discharged. On the other hand, the estimate does not take into account that the several chargings are not completely sequential. The OTA already starts turning on to some extent, when the bias current has not completely settled to its final value yet.

4.6.2 Fast Switching Method

In the reported Switched Opamp realizations, no matter whether the entire OTA is switched [Cro 94] [Pel 97c] or only its output stage [Bas 97a] [Bas 97c], the switching of the amplifiers is done by turning off the bias currents, as explained in the former section. Here a completely new way of switching the amplifiers is developed: it is to interrupt the current path from the power supplies to the circuit with a switch. [Pel 97d][Pel 98b][Pel 98a] This has two distinct advantages:

- The amplifier turn-on time is shorter than in the case of current switching. This is clear from (4.124) derived below.
- All switches are driven by a buffer and therefore their turn-on and turn-off speed is under better control than in the current switching case.

Two types of supply rails are distinguished internally:

$V_{DD,BIAS}$ and $V_{SS,BIAS}$ are the internal supply nodes delivering a constant bias current.

$V_{DD,DYN}$ and $V_{SS,DYN}$ are the internal supply nodes delivering a dynamic current.

These are shown in Fig. 4.19 Both of these have a separate switch. This has several advantages:

- The bias currents are not influenced by the internal dynamic supply voltage which may be changing during the settling process.
- The sources of the current mirrors (M_{2a} and M_{3a} in Fig. 4.11, for example) are tied together, which would not be the case if each branch had a separate switch. Hence the mirroring process is not at all affected by the changing voltage drop over the switch which conducts a dynamic current, or by $V_{GSmirror}$ differences, caused by mismatches in the switches.
- The total (dynamic) switch size is half of the size needed if each branch would have a separate switch. A dynamic current flows each phase, either to one side or the other. A single switch is always conducting, while separate switches would be conducting only one out of two phases on average. In both cases the minimum area requirement determined from the maximum allowable voltage drop and resistance is the same for each switch. In short: the single switch occupies the area of only one of the two switches otherwise needed.

A good estimate for the turn-on time of the low voltage class AB OTA of Fig. 4.11 is

$$T_{on,OTA} \approx max\left\{\frac{V_{Tp}(C_{GS2}+C_{GS3})}{I_B}+3\tau_3, \frac{V_{Tn}(C_{GS5}+C_{GS6})}{I_B}+3\tau_4\right\} \tag{4.125}$$

$$\tau_3 \approx \frac{(C_{GS2}+C_{GS3})}{g_{m2,Q}} \tag{4.126}$$

$$\tau_4 \approx \frac{(C_{GS5}+C_{GS6})}{g_{m5,Q}} \tag{4.127}$$

Comparing (4.124) to (4.120) it can be concluded that the new method of switching is much faster. The reason is the absence of a current source that needs to turn on before the OTA really can start doing so. Further, the absence of a second stage, which inevitably is paired to a large time constant, also enhances the turn-on speed of the low voltage class AB OTA.

In principle all circuits in a Switched Opamp implementation can be switched. This is so for the differential integrating opamps but also for the CMFB amplifiers. Even input stages to comparators can be switched if desired. As a final note it is pointed out that it is good practice not to switch the bias circuits. Compared to the whole system their current drain is minimal. Even more, turning the bias off is exactly what makes the original switching method slow and is what is avoided in the fast method.

4.6.3 The Switch Behavior

In this section some design issues for the switches in the capacitive network around the opamps are given. Given the low supply voltage, the proper operation of the switches requires some attention.

Consider e.g. the OTA of Fig. 4.1 (but the same holds for Fig. 4.11). The switches, which in the Original Switched Opamp Technique are connected with one terminal to V_{REF} have a driving voltage of

$$V_{GSswitch} = V_{GSIN} + V_{DSsat,p} \tag{4.128}$$

The switch is n-type and the input transistor is p-type. This switch has the advantage of a lower V_T and has $V_{DSsat,p}$ more driving voltage, so the condition for putting it in the linear region is automatically fulfilled, if the input transistors aren't biased too deeply into weak inversion. The switch over-drive voltage is

$$V_{ov,switch} = (V_{Tn} - V_{Tp}) + V_{ov,IN} + V_{DSsat,p} \tag{4.129}$$

So ensuring the switches to conduct, is not a problem. Remember that in the Modified Switched Opamp Technique the choice exists to either make the switch over-drive voltage (and the signal swing) larger with respect to the Original Switched Opamp Technique , or to keep the switch over-drive and reduce the supply voltage!

The above derived over-drive voltage only ensures that the switches can be turned on. It does not mean that they are always in the linear region. Indeed, considering the switch over-drive voltage, it happens that a large capacitive feed-forward in the beginning of the integration phase, causes the input switch connected to the opamp input to be in saturation rather than in the linear region. This is so if

$$V_{sw,ov} < V_{ff} \tag{4.130}$$

where V_{ff} is the feed-forward voltage. This means that the switch discharges the sampling capacitor with a constant current. Hence a slew rate can be defined:

$$SR_{sw} = \frac{C_S}{I_{max,sw}} \tag{4.131}$$

$$I_{max,sw} = \frac{KP_{sw}}{2}\left(\frac{W_{sw}}{L_{sw}}\right)(V_{ov,sw})^2 \tag{4.132}$$

The relationship between the switch slew rate and the opamp slew rate must be examined. The current discharging the sampling capacitor flows into the integrating capacitor and is drawn by the opamp. Consider the opamp to have a maximum output current of $I_{max,ota}$, and assume that all of it can be applied to draw the signal current, i.e., there is no significant output capacitance. If the feed-forward voltage is such that the switch is in the saturation regime, there are two possible scenarios.

$$\begin{cases} I_{max,sw} < I_{sl,ota} \implies SR = \dfrac{C_S}{I_{max,sw}} \quad \text{and} \quad V_o = V_{REF} \\ I_{max,sw} > I_{sl,ota} \implies SR = \dfrac{C_S}{I_{max,ota}} \quad \text{and} \quad V_o > V_{REF} \end{cases} \tag{4.133}$$

where V_o is the voltage at the opamp input terminal. In the first case the slew rate is switch limited, in the second it is limited by the opamp. If it is switch limited, it means that the opamp is over-dimensioned, since its current capabilities are never used. This also means that power is being wasted, because the maximum current capability of an opamp is always related to the quiescent currents. On the other hand, the switch sizes should be minimized to avoid the bad influences caused by its parasitic capacitances. Therefore it is best to make the switch size such that the following condition is fulfilled

$$I_{max,sw} = I_{sl,ota} \tag{4.134}$$

Note that if in reality a feed-forward signal with the opposite sign occurs, the switch is not in the saturation regime, but in the linear region, albeit with a very large over-drive voltage. Consequently that condition does not cause a problem.

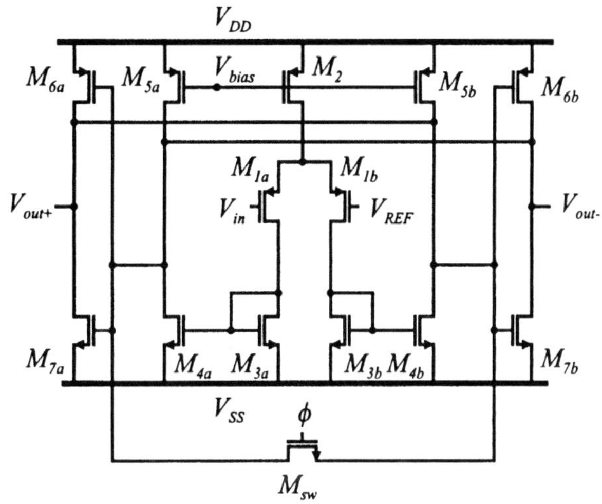

Figure 4.20. *The regenerative comparator used in the realization of Section 6.1.*

4.7 Comparators

Comparators are a necessary component of a $\Delta\Sigma$ modulator. Here the problems related to their operation with low supply voltage are highlighted.

4.7.1 A Regenerative Comparator

In the realization of Section 6.1 the comparator shown in Fig. 4.20 [Ste 88] has been used. It consists of a symmetrical OTA as an input stage to a latch. It is a high speed comparator, but it requires a supply voltage greater than the sum of the the threshold voltages. Here it has been used for a supply voltage of $1.5V$. Consequently, when resetting to the meta-stable point, the inverters are biased in weak inversion. In fact the meta-stable point is not chosen exactly in the middle of the supply voltage, but slightly below. This is necessary because a good reset requires the switch transistor M_{sw} to have low on resistance and thus at least $100\,mV$ over-drive voltage. With a V_{Tn} of $750\ mV$, the meta-stable point is chosen around $600\,mV$.

This is an example of a high voltage circuit being applied to low supply voltage, by pushing the limits. This comparator cannot be used at lower supply than $1.5V$. In the next section a dedicated very low voltage comparator is introduced.

4.7.2 Low Voltage Comparator

For very low supply voltage, it doesn't suffice to force an existing circuit to operate with the low supply. It requires the development of dedicated circuits. It enhances reliability to keep these as simple as possible. In the issue of low voltage comparators, the most difficult to do is the reset. A $\Delta\Sigma$ modulator doesn't need a comparator with high resolution, but it requires low hysteresis [Bos 88], so a good reset is important.

A low voltage comparator requires the following functional items.

input stage A differential stage is about the simplest input stage possible. Low voltage operation of it is well-known from the Switched Opamp

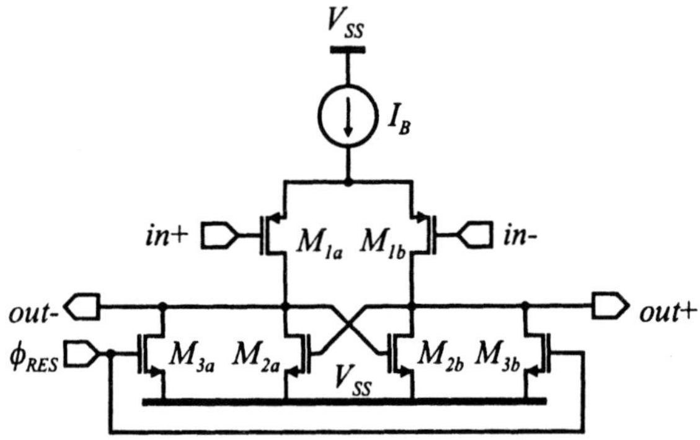

Figure 4.21. *The regenerative comparator used in the realization of Section 6.2. The reset is not done to the meta-stable state, because that level cannot be switched. Instead the reset is done to V_{SS}.*

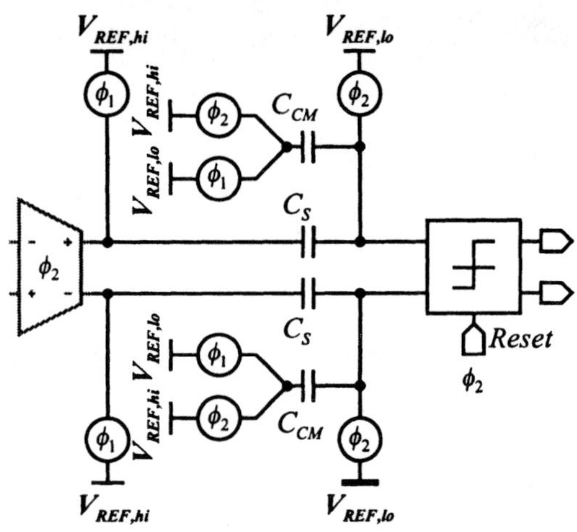

Figure 4.22. *The level shift network shifts the common mode level of the last integrator output signals from $V_{DD}/2$ to V_{SS}, which is the common mode input level of the comparator of Fig. 4.21.*

technique. It is intended to have its common mode input at V_{REF}, i.e., $V_{REF,lo}$ in the Modified Switched Opamp Technique.

regenerative section — Two *n*MOST inverters, in positive feedback configuration satisfy for regeneration, although it is not the fastest possible.

reset — The reset can be done to V_{SS} instead of to the meta-stable state, which is not possible because a level in between the supply voltage cannot be switched in the Switched Opamp technique.

The combination of these requirements leads to the comparator of Fig. 4.21. The operation of this comparator is as follows. The active phase starts when ϕ_{RES} goes low. The switches $M_{3a,b}$ become a high impedance. The output nodes rise quasi simultaneously to the meta-stable point,

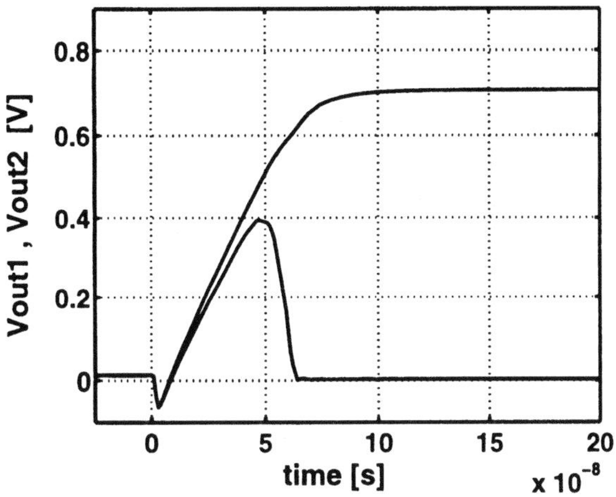

Figure 4.23. *The Low Voltage comparator waveforms. After reset to V_{SS} the output nodes rise quasi simultaneously to the meta-stable point, from which point on the two outputs diverge.*

from which point on the two outputs diverge. The divergence only occurs from the meta-stable point on, because only then the regeneration transistors $M_{2a,b}$ are properly biased and have their full transconductance, which determines the comparison speed. The low output is V_{SS}. The high output doesn't rise completely to V_{DD}, because the full I_B continues to flow through the current source and the input transistor of the high side. When the ϕ_{RES} goes low, the output nodes are shorted to ground and a reset occurs. Fig. 4.23 shows the output signals after reset.

The conversion time is determined by two phenomena. The first is the simultaneous rising of the output voltages to the meta-stable point. The second is the time needed to diverge to the positive and negative supply rail. A linear model for the regenerative comparator for a constant input voltage yields [Roo 96]

$$V_{out}(t) = \left(1 + \frac{g_{min}}{g_{mr}}\right) V_{in} e^{\frac{g_{mr}}{C}t} - \frac{g_{min}}{g_{mr}} V_{in} \quad (4.135)$$

$$g_{mr} = g_{m3a,b} \quad (4.136)$$

$$g_{min} = g_{m1a,b} \quad (4.137)$$

$$C = C_{GS3a} + C_L + C_{DB2a,b} + C_{DB3a,b} \approx C_{GS3a} \quad (4.138)$$

The conversion time can be approximated by

$$T_c \approx \frac{V_{DD}C}{I_B} + \tau_r ln\left(\frac{V_{DD}}{2V_{in,min}} \frac{1}{\left(1 + \frac{g_{min}}{g_{mr}}\right)}\right) \quad (4.139)$$

$$\tau_r = \frac{C}{g_{mr}} \quad (4.140)$$

The first part accounts for the rising to the metastable point. The second part is the exponential settling process. The voltage $V_{in,min}$ is the smallest voltage for which a conversion is required.

Note that g_{mr} and g_{min} are of the same order of magnitude, since they draw the same current and are biased in the moderate inversion region.

As the output of the preceding integrator, sampled to $V_{REF,lo}$ has a DC offset of $(V_{REF,hi}-V_{REF,lo})/2$, a level shift needs to be applied to bring the sample to a DC level of $V_{REF,lo}$. Fig. 4.22 shows a capacitive level shift network. The proper level shift adjust capacitor sizing is

$$C_{CM} = \frac{C_S}{2} \tag{4.141}$$

4.8 Conclusions

In this chapter low voltage circuit design issues are treated. The minimum achievable supply voltage is investigated. A low voltage current mirror is studied. A differential input stage consisting of the addition of one transistor to the mentioned current mirror is introduced. These two blocks are the cornerstones of the low voltage class AB OTA topology especially designed for switched opamp applications, which is presented. A custom CMFB error amplifier based on a "trifferential" input stage is introduced. Also discussed are some specific Switched Opamp design techniques such as switching the opamps. The classic and the newly presented techniques are treated and compared. Finally comparators are treated, because they are an essential part of a ΔΣ modulator. It is shown how a very low voltage comparator can be constructed. In general it is demonstrated that low voltage techniques require dedicated low voltage circuits.

5
DESIGN AND POWER CONSIDERATIONS

In Chapter 3 the main low voltage system building blocks were introduced to build the $\Delta\Sigma$ modulators of Chapter 2. These building blocks are integrators. In Chapter 4 the circuits to implement these with, given the low voltage restriction, and their properties are discussed. In order to be able to proceed to the design phase the required circuit specifications need to be known. It is necessary to know what gain and GBW the OTAs must be designed to have. For that reason this chapter investigates the settling process of a non-ideal Switched Opamp integrator. Finite gain and GBW effects are studied. Some other design issues are also treated.

Secondly the power aspect of the design is treated. The implications of the low supply voltage restriction on the system power consumption is studied.

5.1 Settling Behavior of Integrator

In this section the integrator settling characteristics are studied in view of the main non-ideal circuit properties, such as OTA gain and GBW. Given the models, behavioral simulations can reveal the required specifications in order to achieve the target system performance. The models are derived for the specific case of the Differential Modified Switched Opamp integrator. The starting-point is the charge conservation equation in the charge transfer process that takes place in an integrator. Depending on the model of the active element the following step can include the solution of a differential equation. For this purpose the initial conditions must be known. They depend on the capacitive feed-through signal. This signal is interesting for a second reason. The initial feed-through signal can forward bias parasitic junction diodes, because the reference voltage $V_{REF,lo}$ is at V_{SS}. It must be checked that this effect is acceptable. For these reasons the feed-through is studied first.

5.1.1 Capacitive Feed Through

The capacitive feed-through is caused by immediate charge redistribution when the network, consisting of capacitors and switches, makes the transition between the sampling and integrating configuration. The amplifier is assumed to have a larger time constant than this effect, and thus can be considered to be inactive at that instant.

From Fig. 3.18 and also considering the feedback capacitors shown in Fig. 6.8 the charge conservation equations can be written down as

$$\begin{cases} q_S^- + q_{CM}^- + q_{FB}^- + q_I^- = q_S^+ + q_{CM}^+ + q_{FB}^+ + q_I^+ \\ q_I^- + q_L^- + q_{CMS,eq}^- = q_I^+ + q_L^+ + q_{CMS,eq}^+ \end{cases} \tag{5.1}$$

Substituting for the charges in above expression and making use of the voltage law, an expression for the initial voltages at the amplifier input and output terminals is derived to be:

$$\begin{aligned} V_o^+ = {} & V_{REF,lo} \\ & + \Big[-C_S\, V_{in}^- + (C_S - C_{CM})\,(V_{REF,hi} - V_{REF,lo}) - C_{FB}\, V_{REF,\Delta\Sigma} \\ & + \frac{C_{INT}\, C_L\,(V_{REF,hi} - V_{REF,lo}) - C_{INT}\,(C_L + C_{CMS,eq})\, V_{out}^-}{C_{INT} + C_L + C_{CMS,eq}} \Big] \Big/ \\ & \left(\frac{C_{INT}\,(C_L + C_{CMS,eq})}{C_{INT} + C_L + C_{CMS,eq}} + C_S + C_{CM} + C_{FB} \right) \end{aligned} \tag{5.2}$$

$$\begin{aligned} V_{out}^+ = {} & V_{REF,lo} + \Big[C_{INT}\,(C_S + C_{CM} + C_{FB})\, V_{out}^- \\ & +(C_{INT}\,(C_S + C_L) + C_L\,(C_S - C_{FB}) \\ & +(C_L - C_{INT})\, C_{CM})(V_{REF,hi} - V_{REF,lo}) \\ & - V_{in}^- C_{INT}\, C_S - V_{REF,\Delta\Sigma}\, C_{INT}\, C_{FB} \Big] \Big/ \\ & \Big(C_{INT}\,(C_L + C_{CMS,eq}) + \\ & (C_{INT} + C_L + C_{CMS,eq})(C_S + C_{CM} + C_{FB}) \Big) \end{aligned} \tag{5.3}$$

These expressions are for a positive feedback value. For negative feedback the sign of $V_{REF,\Delta\Sigma}$ inverts. It is V_o^+ which can go below V_{SS}. The worst case is for large input and output voltage and for a positive feedback signal.

The first integrator in the system (as shown in Fig. 6.7) is slightly different due to the absence of the C_{CM} capacitances. The feed-through expressions are now

$$\begin{aligned} V_o^+ &= V_{REF,lo} + \Big[- C_S\, V_{in}^- - C_{FB}\, V_{REF,\Delta\Sigma} \\ &+ \frac{C_{INT}\, C_L\, (V_{REF,hi} - V_{REF,lo}) - C_{INT}\,(C_L + C_{CMS,eq})\, V_{out}^-}{C_{INT} + C_L + C_{CMS,eq}} \Big] \Big/ \\ &\left(\frac{C_{INT}\,(C_L + C_{CMS,eq})}{C_{INT} + C_L + C_{CMS,eq}} + C_S + C_{FB} \right) \end{aligned} \tag{5.4}$$

$$\begin{aligned} V_{out}^+ &= V_{REF,lo} + \Big[C_{INT}\, \big((C_S + C_{FB})\, V_{out}^- - V_{in}^- C_S - V_{REF,\Delta\Sigma}\, C_{FB}\big) \\ &+ (C_L\,(C_S + C_{INT}))(V_{REF,hi} - V_{REF,lo}) \Big] \Big/ \\ &\Big(C_{INT}\,(C_L + C_{CMS,eq}) + (C_{INT} + C_L + C_{CMS,eq})(C_S + C_{FB}) \Big) \end{aligned} \tag{5.5}$$

The same remarks hold here. The worst case capacitive feed-through signals can now be calculated for a specific design.

For a supply voltage of $0.9V$, $V_{REF,hi} = V_{DD} = 0.9V$ and $V_{REF,lo} = V_{SS} = 0V$ some worst case assumptions can be made about maximum input and output signals. For the first type integrator the maximum input signal is limited to $700\,mV$ while the feedback signal $V_{REF,\Delta\Sigma}$ equals $200\,mV$. The maximum output signal can be $700\,mV$. For the first integrator the maximum input signal doesn't include a DC component and is $100\,mV$. The assumptions for the relative capacitance values are $C_s = C_{FB} = 1$ and $2\,C_{CM} = 1$ and $C_L = C_{CMS,eq} = 1$ for the full integrator and 0.6 for the first one. Capacitors C_{INT} are 2 for the full integrator and 5 for the first integrator. The worst case feed-through values are listed in Table 5.1. For the full integrator this is $-129\,mV$ and for the first one it is $-183\,mV$. This means that the parasitic junction diode of a switch diffusion region is being forward biased by this amount. These values are reasonable and do not cause considerable problems. Furthermore, in reality, the feed-through values will be less than this because really worst case parameters have been assumed and also the non-zero resistance of the switches lessens the feed-through value.

int. type	V_{in}^-	V_{FB}	V_{out}^-	$V_{REF,hi}$	$V_{REF,lo}$	$V_{in}^+ +$	V_{out}^+
full	$700\,mV$	$200\,mV$	$700\,mV$	$900\,mV$	$0V$	$-129\,mV$	$386\,mV$
first	$100\,mV$	$200\,mV$	$700\,mV$	$900\,mV$	$0V$	$-183\,mV$	$504\,mV$

Table 5.1. *The worst case capacitive feed-through signals for the* full *and* first *Modified Switched Opamp integrator.*

It can be concluded that the capacitive feed-forward signal can be below V_{SS} but it does not endanger the functionality of the Modified Switched Opamp integrator.

5.1.2 Settling process

The quality of the performance of the overall system is function of the specifications of the individual building blocks. [Bos 88][Rib 91][Fee 91][Fis 97] In the next sections the necessary OTA gain and *GBW* is determined. For this purpose a few models for the integrators are elaborated. Behavioral simulations of the ΔΣ modulator are carried out next. From those the minimal necessary specifications can be derived. First a model is presented taking only finite gain into account. Then a model of an integrator with a certain unity gain frequency is elaborated. The next model takes both finite gain and *GBW* into account. The first models provide an initial value, while the last one gives a more accurate determination of the needed specifications.

5.1.2.1 Finite Gain Effects.

The effect of the finite gain in a control loop is that a static error results. Modeling an integrator with only limited gain effects can be done by considering an opamp for the active element. The following analysis is based on the single ended version of the Differential Modified Switched Opamp integrator of Fig. 3.18. First the charges can be expressed in terms of the voltages across them at the end of the two phases. Phase ϕ_1 is considered to be the sampling phase and ϕ_2 is considered the integration phase.

$$\begin{cases} q_{S1} &= C_S\left(\dfrac{V_{REF,hi}+V_{REF,lo}}{2}+v_{in1}\right) \\ q_{CM1} &= C_{CM}(V_{REF,hi}-V_{REF,lo}) \\ q_{FB1} &= C_{FB}V_{REF,\Delta\Sigma} \\ q_{INT1} &= C_{INT}(V_{OUT}+v_{out,n-1}-v_{o,n-1}) \end{cases}$$
$$\begin{cases} q_{S2} &= C_S(V_{REF,hi}-v_{o,n}) \\ q_{CM2} &= C_{CM}(-V_{o,n}) \\ q_{FB2} &= C_{FB}V_{o,n} \\ q_{INT2} &= C_{INT}(V_{OUT}+v_{out,n}-v_{o,n}) \end{cases} \tag{5.6}$$

Next two system equations can be written down. The first one states that all charge on the capacitors connected to the opamp input is conserved during the integration phase. The second is the opamp equation.

$$q_{S1}+q_{CM1}+q_{FB1}+q_{INT1} = q_{S2}+q_{CM2}+q_{FB2}+q_{INT2} \tag{5.7}$$
$$v_{out,\phi_2} = -Av_{o,\phi_2} \tag{5.8}$$

The system transfer function is then given by

$$v_{out} = \frac{C_S}{C_{INT}}\frac{\alpha z^{-\frac{1}{2}}}{(1-\beta z^{-1})}(v_{in}-V_{REF,\Delta\Sigma}) \tag{5.9}$$
$$\alpha = \frac{1}{\left(1+\dfrac{1}{\mathcal{F}_{dc}A}\right)} \tag{5.10}$$
$$\beta = \frac{\left(1+\dfrac{1}{A}\right)}{\left(1+\dfrac{1}{\mathcal{F}_{dc}A}\right)} \tag{5.11}$$
$$\mathcal{F}_{dc} = \frac{C_{INT}}{C_S+C_{FM}+C_{CM}+C_{INT}} \tag{5.12}$$

These are for a positive feedback signal. It suffices to change the sign of the term in $V_{REF,\Delta\Sigma}$ for negative feedback. The effect of a finite gain on the ideal integrator transfer function is a gain

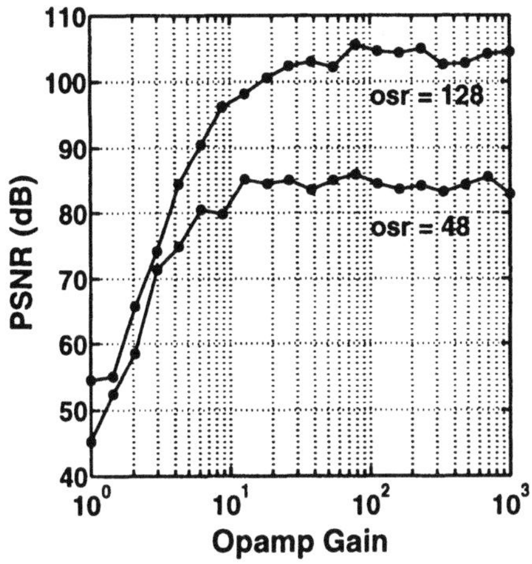

Figure 5.1. *Performance degradation due to finite gain in the 3rd order single loop Switched Opamp $\Delta\Sigma$ modulator. Two curves are shown to illustrate that higher performance requires higher gain. For an oversampling ratio (R_o) of 48 a rather low gain of 20 seems to be sufficient.*

error factor α and a pole error factor β. The single loop $\Delta\Sigma$ modulators are quite insensitive for integrator gain errors, as long as they are constant, which is the case here because of the linear nature of the degradation. Looking at Fig. 2.5 it can indeed be seen that the optimum coefficients may be varied without harm due to the flat maximum of the performance diagram as function of the coefficients. Of course, a too large deviation is not allowable either. The pole error has other implications. It has a similar effect as white noise: a flat noise floor appears. Hence the in-band noise increases and the performance drops. It is plausible that for higher oversampling ratio the requirements are more strict. The performance for a 3rd order single loop Switched Opamp $\Delta\Sigma$ modulator with optimal coefficients [0.2,0.5,0.5] and for relative input amplitude of 0.55, which is the maximum amplitude before overload, is shown in Fig. 5.1 for two different oversampling ratios. Oversampling ratio 48 is shown, because it is of direct importance for the implementation of Section 6.2. For illustration purposes the case of $R_o = 128$ is also shown. It is know that as a rule of thumb single loop $\Delta\Sigma$ modulators require only a gain of the order of the oversampling ratio [Rib 91][Eyn 93a]. Here for $R_o = 48$ a gain of 20 seems to be sufficient. In practice, the gain is better over-dimensioned for several reasons. First, as the output impedance of a transistor is usually not modeled accurately, a discrepancy between real and simulated gain can be as high as a factor of ten. Further, a higher gain helps reducing the harmonic distortion.

5.1.2.2 Finite Bandwidth.

A second property of an amplifier is that it has a finite bandwidth. In order to isolate only bandwidth effects a model with only a pole at DC is used. [Mar 81][Gei 82]

$$H(s) = -\frac{\omega_u}{s} \tag{5.13}$$

The opamp equation now becomes

$$\frac{dv_{out}(t)}{dt} = -\omega_u v_{out}(t) \tag{5.14}$$

The charge conservation equation (5.7) now also still holds together with (5.6). As initial condition $v^{+}_{out,n} = v_{out,n-1}$ is taken. This means that the voltage at the opamp output does not jump immediately. The settling process is now determined by an exponential settling with a certain time constant. Since the settling time is limited to a half period, the charge transfer is incomplete. This makes the integrator transfer function deviate from the ideal. It is now

$$v_{out} = \frac{C_S}{C_{INT}} \frac{\alpha z^{-\frac{1}{2}}}{(1-z^{-1})(1-\beta z^{-1})} v_{in} + \frac{C_{FB}}{C_{INT}} \frac{(1-K\,sign(y))\,z^{-\frac{1}{2}}}{(1-z^{-1})(1-\beta z^{-1})} V_{REF,\Delta\Sigma} \tag{5.15}$$

$$\alpha = 1-K \tag{5.16}$$

$$\beta = K\,\mathcal{F}_{dc} \tag{5.17}$$

$$K = e^{-\omega_u \mathcal{F}_{dc} \frac{T}{2}} \tag{5.18}$$

$$\mathcal{F}_{dc} = \frac{C_{INT}}{C_S + C_{FM} + C_{CM} + C_{INT}} \tag{5.19}$$

T is the sample period and y is the modulator output.

Simulation results show a rather small dependency of the performance on the unity gain frequency location. Therefore, simulation results of only this effect are skipped, to proceed directly to the combined effect of finite gain and finite frequency behavior in the next section.

5.1.2.3 Combined finite gain and bandwidth.

In reality the system performance is affected by both of the above considered non-ideal properties at the same time. It is not straightforward to derive the performance degradation of the combination of effects from each separately. Hence we must derive the transfer again, but now with a different model taking both finite gain and finite frequency response into account. Since the amplifier to be used is an OTA, a voltage controlled current source is the proper model. It is given a transconductance g_m and an output conductance . The OTA model equation is [Rob 92]

$$i_{out,2}(t) = g_m v_{o,2}(t) + v_{out,2}(t) \tag{5.20}$$

Two equations can be written down. They are the nodal equations for the input and output node of the OTA.

$$I_{C_{INT}}(t) + i_{out,2}(t) + I_{C_L}(t) + I_{C_{CMS,eq}}(t) = 0 \tag{5.21}$$

$$I_{C_S}(t) + I_{C_{CM}}(t) + I_{C_{FB}}(t) = I_{C_{INT}}(t) \tag{5.22}$$

In these equations the total current is considered, and not only the AC component. The equations can be rewritten substituting the time derivatives of the charge expressions $\frac{dq_{j2}}{dt}$ from (5.6) for the currents I_{C_k}. All the DC terms vanish and the following equations are obtained

$$\begin{aligned} C_{INT}\left(\frac{dv_{out,2}(t)}{dt} - \frac{dv_{o,2}(t)}{dt}\right) + g_m v_{o,2}(t) + v_{out,2}(t) \\ +(C_L + C_{CMS,eq})\frac{dv_{out,2}(t)}{dt} = 0 \end{aligned} \tag{5.23}$$

$$(C_S + C_{FB} + C_{CM} + C_{INT})\frac{dv_{o,2}(t)}{dt} = C_{INT}\frac{dv_{out,2}(t)}{dt} \tag{5.24}$$

The initial conditions are from (5.2), (5.3), (5.4) and (5.5). The common mode component is omitted, because it is irrelevant for the purpose of performance evaluation.

$$v^-_{o,2} = \frac{\left(-C_S v_{in,n-\frac{1}{2}} - \left(\frac{C_{INT}C_L}{C_{INT}+C_L}\right) v_{out,n-1}\right)}{\left(\frac{C_{INT}C_L}{C_{INT}+C_L}\right) + C_S} \tag{5.25}$$

$$v^-_{out,2} = \frac{\left(-\left(\frac{C_{INT}C_S}{C_{INT}+C_L}\right) v_{o,n-\frac{1}{2}} + \left(\frac{C_{INT}C_S}{C_{INT}+C_L}\right) v_{out,n-1}\right)}{\left(\frac{C_{INT}C_L}{C_{INT}+C_L}\right) + C_S} \tag{5.26}$$

Solving the differential equations with these initial conditions one finds the transfer functions

$$H(z) = \frac{C_S}{C_{INT}} \frac{\alpha z^{-\frac{1}{2}}}{(1-\beta z^{-1})} \tag{5.27}$$

$$\alpha = R_1(1 - K(r_L + r_S\frac{C_{INT}}{C_S}) - \frac{K}{A} r_S \frac{C_{INT}}{C_S}) \tag{5.28}$$

$$\beta = R_1(1 - K(r_L + \frac{r_S}{A})) \tag{5.29}$$

$$R_1 = \frac{1}{1 + \frac{1}{A\mathcal{F}_{dc}}} \tag{5.30}$$

$$K = e^{-\frac{g_{meff}}{C_{L,eff}}\frac{T}{2}} \tag{5.31}$$

$$g_{meff} = g_m R_1 \tag{5.32}$$

$$C_{L,eff} = C_S + C_{CM} + C_{FB} + \frac{C_L + C_{CMS,eq}}{\mathcal{F}_{dc}} \tag{5.33}$$

$$\mathcal{F}_{dc} = \frac{C_{INT}}{C_S + C_{FM} + C_{CM} + C_{INT}} \tag{5.34}$$

$$r_L = \frac{C_L}{\mathcal{F}_{dc} C_{L,eff}} \tag{5.35}$$

$$r_S = \frac{C_S}{\mathcal{F}_{dc} C_{L,eff}} \tag{5.36}$$

The integrator gain and pole error coefficients α and β are now more involved, although the same effects still play a role. The gain error, which has a similar effect as a coefficient mismatch does not have such a severe implication for a single loop modulator, as long as it is not excessively large. The pole error makes the noise shaping flatten out, and hence directly causes the performance degradation. The expressions show how the effects combine and are indeed not just a simple sum.

Since there are two parameters in this analysis, the gain and transconductance, both could be varied to obtain a three dimensional performance surface. In reality it is much more practical to obtain an estimate for the gain from the analysis in Section 5.1.2.1. Then this gain value can be fixed for the performance assessment varying the transconductance. This type of graph is shown in Fig. 5.2. The topology is also the 3rd order single loop Switched Opamp $\Delta\Sigma$ modulator with optimal parameters [0.2,0.5,0.5]. The relative amplitude of 0.55 is just below overload. The

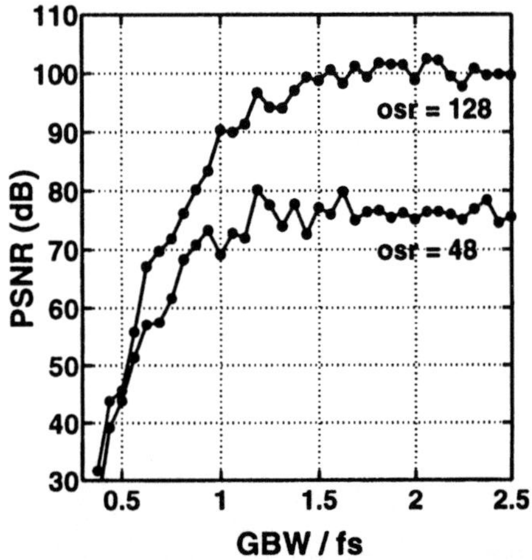

Figure 5.2. *Performance degradation due to finite gain and gain-bandwidth in the 3rd order single loop Switched Opamp ΔΣ modulator. The gain is kept fixed to a value of 200. For oversampling ratio (R_o) of 48 a rather low ratio of the GBW to the sampling frequency of 1.5 is found to be sufficient.*

gain is kept fixed to a value of 200, which is ten times more strict than really necessary in the case of R_o =48 and about 2 times more strict in the case of R_o =128. The capacitor values used to simulate are the ones from the realization in Section 6.2. The effective load capacitance C_{Leff} is about 16.6 pF. Although the g_m is varied, the axis is normalized to GBW relative to the sampling frequency of 1.536 MHz. For the case of an oversampling ratio of 48, a ratio of 1.5 proves to be sufficient.

In reality the integrator has more poles than taken into account by this analysis. It can often be considered a second order system. This means that by putting the non-dominant pole at four times the loop GBW (corresponding to a phase margin of 76°) a critically damped system is obtained. Such a system yields the fastest settling possible, about twice as fast as a first order system. This would mean that the abtained GBW value is over dimensioned. In reality, however, it is not possible to ensure an exact ratio of poles or the phase margin. The increase in settling speed is therefore less. Nevertheless by designing the integrator as a critically damped second order system, an additional safety margin is built in.

5.2 Functionality of CMFB

From (5.2) (5.3) (5.4) (5.5), which express the capacitive feed-through signals right after the transition of the capacitor-switch network from sampling to integrating phase, the behavior of the common mode component of the output signals can also be derived. For the full Modified Switched Opamp integrator, input and output signal are superimposed on a DC (or common mode) value of $(V_{REF,hi} - V_{REF,lo})/2$.

$$V_{in} = \frac{V_{REF,hi} - V_{REF,lo}}{2} + v_{in} \tag{5.37}$$

$$V_{out} = \frac{V_{REF,hi} - V_{REF,lo}}{2} + v_{out} \tag{5.38}$$

Under the condition of proper choices of the capacitors, the feed-forward common mode values have remarkable properties.

$$C_{CM} = \frac{C_S}{2}\,,\ C_{CMS,eq} = C_L \qquad \Longrightarrow \qquad V^+_{in,cm} = V_{REF,lo} \tag{5.39}$$

The Modified Switched Opamp integrator used as the first integrator in the system does not have a DC value on its sampled input, and $V^+_{in,cm} = V_{REF,lo}$ also holds. Especially important is the fact that the charge on C_L which is fully loaded to $(V_{REF,hi} - V_{REF,lo})$ at the end of the sampling phase redistributes with the empty capacitor $C_{CMS,eq}$. The output common mode feed-through signal can be verified to be

$$\{C_{CM} = \frac{C_S}{2}\,,\ C_{FB} = C_S\,,\ C_{CMS,eq} = C_L\} \Longrightarrow V^+_{out,cm} = \frac{V_{REF,hi} - V_{REF,lo}}{2} \tag{5.40}$$

It can be concluded that the feed-forward common mode signals have the proper value by nature, if the capacitor sizes are correct and if the redistribution takes fully place. In principle this means that once in stationary state no significant action is demanded of the CMFB control system.

The analysis of a Modified Switched Opamp integrator cell assumes the presence of a DC voltage drop over the integrating capacitor. This situation however does not exist at startup. When the system is not powered, all capacitors are discharged through leakage currents, however small they are. It must thus be verified that upon startup the voltage drop is built up. For this purpose the charge redistribution as function of a non-zero opamp input voltage $V_{o,n}$ can be studied. Looking at Fig. 3.18 the charge conservation equation can be written for the capacitors connected to the opamp input for a single side, considering only common mode voltages and a common mode voltage error and referring the voltages to $V_{REF,lo}$:

$$q_{S_{n-\frac{1}{2}}} + q_{CM_{n-\frac{1}{2}}} + q_{I_{n-1}} = q_{S_n} + q_{CM_n} + q_{I_n} \tag{5.41}$$

$$\begin{aligned}
&(V_{CM,in} + \Delta V_{CM,in})C_S + V_{REF,hi}C_{CM} \\
&+(V_{CM,out} + \Delta V_{CM,out} - V_{o,n-1})C_{INT} \\
&= \\
&(V_{REF,hi} - V_{o,n-1})C_S - V_{o,n-1}C_{CM} + \\
&(V_{CM,out,n} + \Delta V_{CM,out} - V_{o,n})C_{INT}
\end{aligned} \tag{5.42}$$

In this equation the CMFB action is apparent by the appearance of $\Delta V_{CM,out}$ by which the common mode at the output is kept fixed. This yields

$$V_{o,n} = \frac{C_{INT}}{C_S + C_{CM} + C_{INT}} V_{o,n-1} + \frac{C_S}{C_S + C_{CM} + C_{INT}} \Delta V_{CM,in} \tag{5.43}$$

So a constant output common mode error does not influence the input common mode. This expression can be rewritten as a function of the initial $V_{o,0}$ and a constant $\Delta V_{CM,in}$ as

$$\begin{aligned}
V_{o,n} = \;& \left(\frac{C_{INT}}{C_S + C_{CM} + C_{INT}}\right)^{n+1} V_{o,0} \\
& + \sum_{i=1}^{n} \left(\frac{C_{INT}}{C_S + C_{CM} + C_{INT}}\right)^{i} \left(\frac{C_S}{C_S + C_{CM} + C_{INT}}\right) \Delta V_{CM,in}
\end{aligned} \tag{5.44}$$

Taking the limit of this yields

$$\lim_{n\to\infty} V_{o,n} = 0 \cdot V_{o,0} + 0 \cdot \Delta V_{CM,out} + \left(\frac{C_S}{C_S + C_{CM}} \Delta V_{CM,in}\right) = \frac{2}{3} \Delta V_{CM,in} \tag{5.45}$$

This expression reveals that a constant input common mode error evolves to establish a reduced common mode error on V_o. More important, however, it shows that an initial error of V_o itself is completely removed. In other words the common mode voltage drop over the integrating capacitor is built up starting from any initial condition.

The settling behavior of the CMFB is determined by the CMFB error amplifier's loop GBW. It is treated in Section 4.5.3

5.3 Optimization of the Operating Point

In this section the influence of the operating point on the power consumption of a Switched Capacitor integrator is investigated. For this purpose a differential pair is considered and an expression for the required current to settle to a certain accuracy is derived. In the analysis it is assumed that the gate source capacitance of the input pair can be neglected. In principle it affects the effective load capacitance, due to the feedback configuration. The assumption can be met in the context of very low voltage systems, that are not particularly pushed for speed of operation. In that case the sampling capacitors are very large to have small noise.

5.3.1 Linear Settling

A differential pair is assumed to be biased with a total current of I_B. It has a transconductance of $g_m = \dfrac{I_B}{V_{GS} - V_T}$. From the expression for the settling time as a function of the settling error ϵ of a first order system, the required bias current normalized to the settling period T_s and the load capacitance C can be deduced. When the input stage is biased in strong inversion it is:

$$I_B \frac{T_s}{C} = \frac{1}{\mathcal{F}_{dc}} (V_{GS} - V_T) ln \left(\frac{1}{\epsilon} \right) \tag{5.46}$$

For weak inversion the expression is

$$I_B \frac{T_s}{C} = \frac{1}{\mathcal{F}_{dc}} 2nU_t ln \left(\frac{1}{\epsilon} \right) \tag{5.47}$$

Assuming an under damped second order system with damping factor ζ the settling expression becomes

$$I_B \frac{T_s}{C} = \frac{1}{\mathcal{F}_{dc}} \frac{1}{2\zeta} (V_{GS} - V_T) \left[ln \left(\frac{1}{\epsilon} \right) - \frac{1}{2} \ln(1 - \zeta^2) \right] \tag{5.48}$$

for a bias in strong inversion and

$$I_B \frac{T_s}{C} = \frac{1}{\mathcal{F}_{dc}} \frac{1}{2\zeta} 2nU_t \left[ln \left(\frac{1}{\epsilon} \right) - \frac{1}{2} \ln(1 - \zeta^2) \right] \tag{5.49}$$

for weak inversion In the next section slewing is added.

5.3.2 Linear Settling with Slewing

An important property of a differential pair with an important influence on the settling behavior is the slew rate. This makes the output voltage of the system (an integrator) settle with a constant slope, because a constant output current is delivered. Eventually the system enters the linear region and the further settling behavior continues with an exponential decay.

5.3.2.1 Strong Inversion.

The transition from slewing to the linear region is gradual. A simple model for a differential pair in strong inversion shows slewing to occur for $v_{in,diff} > V_{GS} - V_T$. The required currents to settle to an accuracy of ϵ normalized to T_s and C are

$$I_B \frac{T_s}{C} = V_{step} - (V_{GS} - V_T) + \frac{1}{\mathcal{F}_{dc}}(V_{GS} - V_T) ln\left(\frac{\rho}{\epsilon}\right) \tag{5.50}$$

for the first order case and

$$I_B \frac{T_s}{C} = V_{step} - (V_{GS} - V_T) + \frac{1}{\mathcal{F}_{dc}}\frac{1}{2\zeta}(V_{GS} - V_T)\left[ln\left(\frac{\rho}{\epsilon}\right) - \frac{1}{2}\ln(1-\zeta^2)\right] \tag{5.51}$$

for the second order case, and $\rho = \dfrac{V_{GS} - V_T}{V_{step}}$. The quantity V_{step} is the step input voltage.

5.3.2.2 Weak Inversion.

A simple model for a differential pair in weak inversion considers slewing to occur for $v_{in,diff} > 3nU_t$. The required currents to settle to an accuracy of ϵ normalized to T_s and C are

$$I_B \frac{T_s}{C} = V_{step} - 3nU_t + 2nU_t \frac{1}{\mathcal{F}_{dc}} ln\left(\frac{\rho}{\epsilon}\right) \tag{5.52}$$

for the first order case and

$$I_B \frac{T_s}{C} = V_{step} - 3nU_t + 2nU_t \frac{1}{\mathcal{F}_{dc}}\frac{1}{2\zeta}\left[ln\left(\frac{\rho}{\epsilon}\right) - \frac{1}{2}\ln(1-\zeta^2)\right] \tag{5.53}$$

for the second order case, and $\rho = \dfrac{U_t}{2\,V_{step}}$. The quantity V_{step} is the step input voltage.

5.3.3 Comparison

In Fig. 5.3(a) the required current is shown, normalized to the settling time T_s and the load capacitance C, to settle to an accuracy of 0.1%, 0.01% and 0.001% and as function of the step input voltage. The following values are used in the calculation: weak inversion slope $n = 1.5$, feedback factor $\mathcal{F}_{dc} = 1$, damping factor $\zeta = 0.82$. The dash-dot line is for a differential pair with $V_{GS} - V_T$ of 200 mV. The solid line is for a differential pair biased in weak inversion. For increasing step size the power increases, but more important is that a bias in strong inversion consumes more power than in weak inversion in order to reach the same settling performance. This is clear from Fig. 5.3(b), where the ratios of the curves in (a) are plotted.

It can be concluded that for low power consumption a bias in weak inversion is to be chosen. In practice this means a bias in the transition between weak and strong inversion. In that zone the transistor characteristic in not perfectly exponential as is assumed in the model which is deployed in the above derivation. However, the parasitic capacitors are much smaller than in the case of deep weak inversion.

5.4 Noise

While the performance of the theoretical $\Delta\Sigma$ modulator is only determined by the in-band quantization noise suppression, the physical implementation of the system with solid state devices has to deal with other noise sources as well. The additional noise of a CMOS implementation is thermal noise and 1/f noise. The noise sources are the switches, the opamps and the voltage

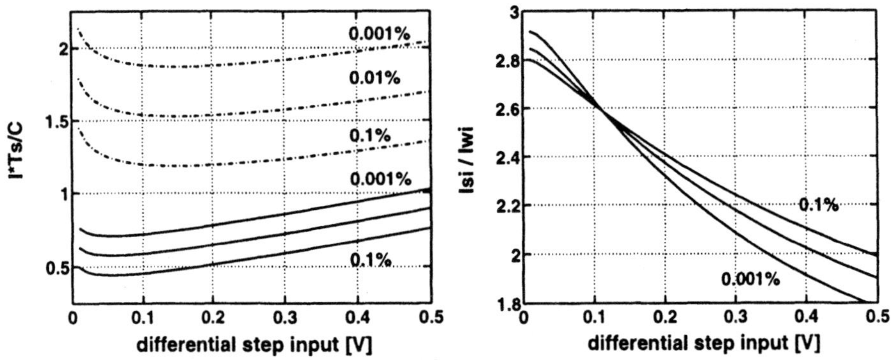

(a) (b)

Figure 5.3. *In (a) the required current is shown, normalized to the settling time T_s and the load capacitance C, to settle to an accuracy of 0.1%, 0.01% and 0.001% and as function of the step input voltage. The dash-dot line is for a differential pair with $V_{GS} - V_T$ of 200 mV. The solid line is for a differential pair biased in weak inversion. The ratios of the curves from (a) are shown in (b) and show that biasing in strong inversion requires more current than in weak inversion in order to reach the same settling performance.*

reference. There are two noise propagation mechanisms taking place in a Switched Capacitor integrator: there is direct or broadband noise and sampled noise [Gob 80][Fis 82][Gob 83]. The sampled noise is the noise voltage present on the sampling capacitor at the end of the sampling and integration phase. It is caused by the switches during the sampling phase and during the integration phase it is caused by the switches and by the opamp noise. The several same noise sources also generate direct or broadband noise at the integrator's output. The latter can however be neglected in a ΔΣ modulator since the noise of the following stages is suppressed by the gain of the preceding integrators. For the same reason the sampled noise of the integrators in the loop can be neglected. This fact is known, but it hasn't been quantified yet. Therefore an analysis quantifying the suppression of each internal noise source is presented in this section. But first the overall noise expressions are stated. Since such analyses exists, no in depth analysis is given [Yin 94a].

5.4.1 Expressions for Circuit Noise Contributions in a ΔΣ modulator

The total circuit noise taken into account is the noise from the switches and noise from the opamp in the first integrator.

$$P_{N,C,t} = P_{N,sw} + P_{N,amp} \tag{5.54}$$

The noise from the switches is the sum of the sampled noise in the sampling phase and in the integration phase. It is band limited by the RC time constant during the sampling phase and by a combination of this and the amplifier's time constant during the integration phase. The amplifier's influence can with a good approximation be neglected such that each phase a sampled noise of $\frac{kT}{C}$ is added. There is a noise contribution associated with (not created by) each capacitor in parallel with the sampling capacitor. The folowing expression holds for the case of the Modified Switched Opamp integrator (Fig. 6.7 and Fig. 6.8 taken together), which besides the sampling capacitor C_S and the C_{FB} capacitor also has a common mode adjust capacitor C_{CM}.

$$P_{N,sw} = P_{N,sw,samp} + P_{N,sw,int} \tag{5.55}$$

$$= \frac{2kT}{C_S R_o}\left(1 + \frac{C_S}{C_{FB}} + \frac{C_S}{C_{CM}}\right) \tag{5.56}$$

The opamp noise consists of white and $1/f$ noise. The opamp's equivalent input noise must be referred to the sampling capacitor, which explains the factor in the folowing expression for the noise power due to the amplifier.

$$P_{N,amp} = \left(\frac{C_{INT}}{\mathcal{F}_{dc} C_S}\right)^2 \left(\frac{2kT}{3C_{L,eff} R_o}\gamma_w + \gamma_f \frac{KF\, I_{DS}^{AF}}{C_{ox} g_m^2 L_{eff}^2} \ln\left(\frac{f_c}{200\ Hz}\right)\right)$$

$$\gamma_x = \gamma_{x,OTA} + \gamma_{x,CMFB} \tag{5.57}$$

The band limiting of the amplifier's white noise is determined by its effective load capacitance $C_{L,eff}$. Both noise terms contain excess noise factors γ_w and γ_f, which take into account the excess noise generated by the other transistors than the input transistors, of the differential amplifier itself and its CMFB amplifier, in the case of a differential implementation. Note that integrators input referred noise due to the OTA is strongly influenced by the integrator gain coefficient C_{INT}/C_S and the feedback factor $\mathcal{F}_{dc}$.

5.4.2 *Suppression of Noise Generated Inside the Loop*

Circuit noise is generated in all the integrators of a $\Delta\Sigma$ modulator. The noise at the first summation point is superimposed on the input signal. It cannot be distinguished from it and hence limits the obtainable performance. The noise injected into the summation points deeper in the loop, are suppressed by the transfer function of the preceding integrators. They have very high gain in the signal band and thus these noise sources are suppressed. In fact they are shaped in a way similar to the quantization noise. Since the noise inside the loop is suppressed, larger noise is allowed. Thus smaller capacitance can be used for an inner loop integrator. The amplifier then drives a smaller load and consequently power can be saved. This phenomenon is known, but the amount of suppression has not been quantified yet [Eyn 93b]. The next analysis presents the factors with which white noise $P_{N,k}$ injected into summing node k, is suppressed.

The noise sources $\overline{dv_k^2}$ are assumed to be white noise sources at summing node k, where $k = 0$ is the input summing node and k in general is the output node of k_{th} integrator. The transfer from node k to the output is

$$H_k = \frac{\prod_{j=k+1}^{n} a_j k_Q \left(\frac{z^{-\frac{1}{2}}}{1+z^{-1}}\right)^{n-k}}{1 + \sum_{i=0}^{n}\prod_{j=i}^{n} a_j k_Q \left(\frac{z^{-\frac{1}{2}}}{1+z^{-1}}\right)^{k-i+1}} \tag{5.58}$$

$$|H_k(z)| \approx \frac{\left|1 - z^{-1}\right|^k}{\prod_{i=0}^{k} a_i} \tag{5.59}$$

in which $a_0 = 1$ by definition and k_Q is the quantizer gain as in Section 2.3.2. The z to frequency transform $z = e^{\frac{j2\pi f}{fs}}$ is applied and yields

$$|H_k(f)| \approx \frac{\left|1 - e^{-\frac{j2\pi f}{fs}}\right|^k}{\prod_{i=0}^{k} a_i} \tag{5.60}$$

$$= \frac{\left|e^{-\frac{j2\pi f}{2fs}}\right|^k \left|e^{\frac{j2\pi f}{2fs}} - e^{-\frac{j2\pi f}{2fs}}\right|^k}{\prod_{i=0}^{k} a_i} \tag{5.61}$$

$$\approx \frac{\left|(2\sin\left(\frac{\pi f}{fs}\right))\right|^k}{\prod_{i=0}^{k} a_i} \tag{5.62}$$

$$\approx \frac{\left|\left(2\frac{\pi f}{fs}\right)\right|^k}{\prod_{i=0}^{k} a_i} \tag{5.63}$$

where both the last approximations hold for $f << f_S$. The equivalent input noise power from the noise sources in the loop is

$$P_{N,eq,in} = \frac{\sum_{k=0}^{n} P_{N,out,k}}{|H_x|^2} = \sum_{k=0}^{n} P_{N,out,k} \tag{5.64}$$

because $|H_x|^2 = 1$ in the signal band ((2.20)). The noise at the output generated at the k_{th} summing point is

$$P_{N,out,k} = 2\int_0^{BW_S} \overline{dv_{k,samp}^2}\, |H_k(f)|^2\, df \tag{5.65}$$

The sampled noise power spectral density equals

$$\overline{dv_{k,samp}^2} = \overline{dv_k^2}\,\frac{BW_{N,k}}{fs} = \frac{P_{N,k}}{fs} \tag{5.66}$$

This is illustrated by Fig. 5.4, along with the shaping which the noise from the several origins undergo.

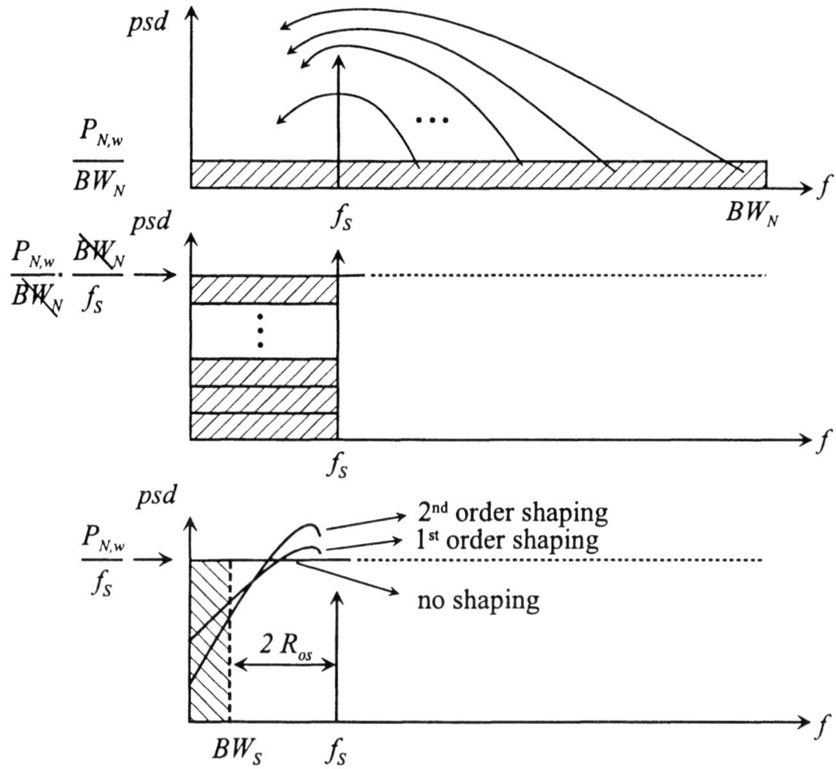

Figure 5.4. *Illustration of sampled noise aliasing. Sampled white noise is aliased into the baseband and piled up a number of times. It appears at the output in a shaped form, depending on where in the loop it is generated.*

Thus

$$P_{N,out,k} = 2\int_0^{BW_S} \frac{P_{N,k}}{f_S} |H_k(f)|^2 \, df \tag{5.67}$$

$$= 2\frac{P_{N,k}}{f_S}\int_0^{BW_S} \left(\frac{\left|2\frac{\pi f}{f_S}\right|^k}{\prod_{i=0}^{k} a_i}\right)^2 df \tag{5.68}$$

$$= P_{N,k}\frac{\pi^{2k}}{(2k+1)\left(\prod_{i=0}^{k} a_i\right)^2}\left(\frac{1}{R_o}\right)^{2k+1} \tag{5.69}$$

Thus the equivalent input noise is

$$P_{N,eq,in} = \sum_{k=0}^{n} P_{N,k}\frac{\pi^{2k}}{(2k+1)\left(\prod_{i=0}^{k} a_i\right)^2}\left(\frac{1}{R_o}\right)^{2k+1} \tag{5.70}$$

The circuit noise generated inside the loop is thus suppressed with the $2k+1$ power of R_o. For $k = 0$ the suppression is $\frac{1}{R_o}$, which is the known benefit of oversampling. For $k = 1$ the

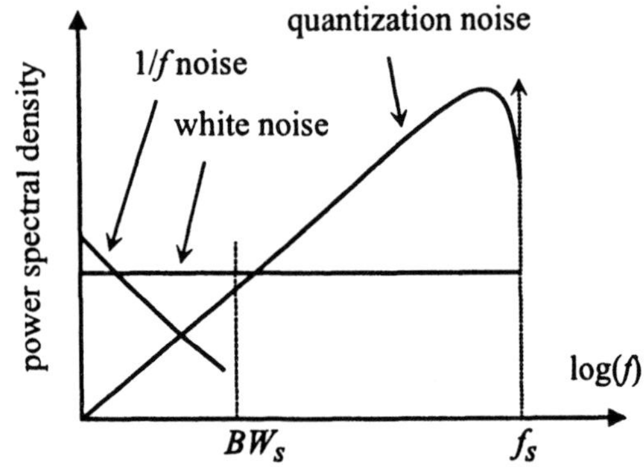

Figure 5.5. *Besides the quantization noise, also the circuit noise sources have to be taken into account. These are white noise from the switch resistances and from the amplifier, and $1/f$ noise from the amplifier*

suppression is $\frac{\pi^2}{3a_1R_o^{\,3}}$. The factor by which the noise at the second summing node is suppressed relatively to that at the first summing node is thus

$$F_{sup,1} = \frac{3a_1^2R_o^{\,2}}{\pi^2} \tag{5.71}$$

and in general for the kth node

$$F_{sup,k} = \frac{R_o^{\,2k}}{\pi^{2k}}(2k+1)\left(\prod_{i=0}^{k} a_i\right)^2 \tag{5.72}$$

This fact allows the integrators in the loop to be much down scaled. In the realization of Section 6.2 the second integrator sampling capacitor has been scaled by a factor of 10.

5.4.3 Noise Trade-off

The equivalent input quantization noise level is related to the reference voltage $V_{REF,\Delta\Sigma}$. The integrated in-band quantization noise can e.g. be calculated by dividing the a non-overloading input RMS voltage by the corresponding ideal *SNR*. A schematic representation of the several noise levels is shown in Fig. 5.5. The *SNR* of a ΔΣ modulator is determined by the ratio of the signal power P_S and both the circuit and the quantization noise ($P_{N,C}$ and $P_{N,Q}$, respectively):

$$SNR = \frac{P_S}{P_{N,C} + P_{N,Q}} \tag{5.73}$$

The designer has the freedom to choose each of these noise contributions separately. Diminishing the quantization noise can be done by choosing a higher oversampling ratio R_o or a higher modulator order. Diminishing the thermal noise is done by choosing a larger sampling capacitor and a larger oversampling ratio. All of these operations require an extra power allocation. A higher oversampling ratio requires all the integrators to settle faster: slew rates and gain bandwidth products must increase. The latter properties are proportional to the bias current. Therefore, in a first order estimation, the extra power consumption is equal to the relative R_o

increase: doubling R_o causes double power consumption. A higher modulator order requires one extra integrator (read opamp) for each additional order in classic Switched Capacitor and Modified Switched Opamp circuits and two in the Original Switched Opamp case. The extra comparator in a cascaded implementation also consumes extra power, but here it is assumed to be much less than that of the integrators taken together. The additional power consumption per integrator in this operation is a relatively small fraction of the total power consumption of all the integrators. This is caused by the following reasons. As proven in the previous section the thermal noise sources internal to the loop are strongly suppressed. Therefore the size of the capacitors inside the loop can be made much smaller than the input sampling capacitors. The corresponding opamps need lower transconductance, which is proportional to the bias current of the input transistors of the OTAs. The entire opamp is scaled down and with it its power consumption. The first integrator needs to drive the input sampling capacitors which are then much larger than the internal capacitors and consequently consume a great portion of the power budget. In order to reduce the thermal noise power by a factor of two, the input sampling capacitor must be increased with a factor of two. The first opamp's power consumption is then also doubled. In order to understand better how the trade-off between quantization- and circuit noise should be done, it is necessary to visualize what *SNR* increase is obtained from a $\Delta\Sigma$ modulator of one order higher. The ideal *SNR* is given by (2.11). The *SNR* increase (or quantization noise suppression) factor going from an n to $n+1$ order single loop modulator is then readily calculated from (2.16), (2.25) and (2.36) to be

$$R_Q = \left(\frac{\Delta_{n+1}}{\Delta_n}\right)^2 \left(\frac{2n+3}{2n+1}\right) \left(\frac{R_o}{\pi}\right)^2 \tag{5.74}$$

$$\Delta_n = \begin{cases} \Delta_n = \prod_{i=1}^{n-1} a_{n,i} a_{n,eff} & \text{for a single loop modulator} \\ \Delta_n = \prod_{i=1}^{m} c_i & \text{for a cascade modulator} \end{cases} \tag{5.75}$$

where $a_{n,i}$ is the ith parameter of an nth order $\Delta\Sigma$ modulator and m the number of cascade stages. For a 2nd to 3rd order transition this gives an extra quantization noise suppression of $14\ dB$ for $R_o = 64$. Assuming that the in-band quantization and circuit noise have equal powers, after adding an order, even in the single loop case the noise power becomes negligible. This corresponds to an effective total in-band noise reduction of $3\ dB$. To continue, an estimate of power consumption should be made.

Based on the numbers obtained from the realization of Section 6.2 and for a 2nd order modulator the first integrator is assumed to consume 70% of the power, and the second one and the comparator each 15%. Similar numbers are also found in [Rab 97] and in [Zwa 96a]. In this last realization e.g., which is a continuous time implementation o 4th order, the first integrator comsumes 67% of the total power, about 8.5% per other integrator and 7.5% for the comparator. A 3rd order modulator is assumed to have the 2nd and 3rd integrator of equal size. Based on these assumptions it is possible to calculate for both the 2nd and 3rd order $\Delta\Sigma$ modulator the net power savings of an additional order for single loop configuration. As pointed out above, adding one order can make the quantization noise negligible compared to the circuit noise. Consequently, for a constant performance (read constant total in-band noise power) the latter can be a allowed to be double. This causes a first integrator power consumption decrease of a factor of two, which is beneficial since it has a large power drain. Table 5.2 shows for both cases the relative power consumption of each integrator, the total power consumption of the new modulator of one order higher and the net power savings in the last column. The power savings are in the order of 20%.

It can be concluded that for low power considerations the in-band quantization noise must never exceed the in-band thermal noise. The in-band quantization noise must be comparable or preferably lower than the in-band thermal noise.

	Relative P Consumption				
case	1st int.	2nd int.	3rd int.	comp.	Net P Saving
$2^{nd} \rightarrow 3^{rd}$ order	70%	15%	–	15%	20%
$3^{rd} \rightarrow 4^{th}$ order	60%	13%	13%	13%	17%

Table 5.2. *In the assumption that the quantization and circuit noise contributions are equal, the addition of an extra order to a ΔΣ modulator makes the quantization noise negligible compared to the circuit noise. This allows a net power saving and indicates that the performance may not be quantization noise limited.*

5.5 Power Considerations

In this section the power aspect of the design of ΔΣ modulators is treated. Low power is a controversial subject. In fact it is not straightforward to define what *low power* operation is. It is plausible that a circuit with a higher frequency of operation requires a higher power budget. It is also plausible that performing analog signal processing with increased accuracy, requires increased power consumption. So, if a certain performance requires a certain power consumption, altering the performance through a redesign should change the necessary power consumption. The proper way to think about low power consumption is to define it as a trade off between contradictory specifications, such as accuracy, frequency of operation or signal bandwidth and power consumption. Further the power supply plays a role that at first looks at least ambiguous. Lowering the power supply at first looks like it lowers the power consumption because the product of voltage and current is smaller. However, lowering the power supply voltage has a number of consequences that on the contrary cause the power consumption to increase, as is shown and illustrated further on. First an expression clarifying the mentioned trade-off is given. Then all the influences are listed and practical advice given.

5.5.1 Expression for Switched Capacitor integrator power consumption

The effects of thermal noise on the speed and power consumption of a one pole analog circuit are discussed in [Vit 90]. That analysis assumes an infinite efficiency of the active part of the circuit, and hence only considers the power that is dissipated when charging and discharging the functional capacitor in the circuit. The assumption of no power consumption in the active circuitry yields a power consumption orders of magnitude below real implementations. Taking into account a simple model for the active parts, can lead to a more realistic bandwidth-resolution-power trade-off.

The analysis here is similar to [Mar 98a]. In that analysis the starting point is not so much the thermal noise, which comes into account later, but the settling process. The simplest approach of a settling process is an exponential settling with a time constant τ, which has a time dependent settling error ϵ_s of

$$\epsilon_s = e^{-\frac{t}{\tau}} = e^{-N_\tau} \tag{5.76}$$

In the above equation the ratio of the time and the time-constant is denoted the number of necessary time constants. For B bits of resolution the settling error must be $\epsilon_s = 2^{(-B)}$. It follows that the number of necessary time constants equals

$$N_\tau = \ln(2^B) = B\ln(2) \tag{5.77}$$

The available time to settle during the integration phase in a Switched Capacitor or Switched Opamp integrator is half a clock period or $T/2$, hence

$$\frac{T}{2} = N_\tau \, \tau \tag{5.78}$$

On the other hand the time constant is determined by the transconductance g_m of the amplifier used in the integrator and the effective load capacitance $C_{L,eff}$:

$$\tau = \frac{C_{L,eff}}{g_m} \tag{5.79}$$

Taking together (5.77),(5.78) and (5.79) the following can be obtained

$$g_{meff} = B \ln(2) C_{L,eff} 2fs \tag{5.80}$$

Capacitance $C_{L,eff}$ is dependent on all the capacitors in the network as e.g., in (5.33). In general it can always be expressed as capacitance C_S of the sampling capacitor multiplied by an excess capacitance factor N_C, such that

$$C_{L,eff} = N_C \, C_S \tag{5.81}$$

The size of C_S is determined by the requirement for the integrated kT/C noise power

$$P_N = N_N \frac{kT}{C_S} \quad \text{or} \quad N_N \frac{kT}{C_S R_o} \tag{5.82}$$

where the latter holds when an oversampling factor of R_o is applied. N_N is a noise excess factor, which can take into account other noise sources, such as the noise related to the switches of extra capacitors and to the amplifier. For a MOS transistor the transconductance can be expressed as

$$g_m = \frac{2I}{V_{ov}} \quad \text{with} \quad V_{ov} = \begin{cases} V_{GS} - V_T & \text{in strong inversion} \\ 2nU_t & \text{in weak inversion} \end{cases} \tag{5.83}$$

The maximum signal power for a sinusoidal signal can be expressed as

$$P_S = \frac{V_S^2}{2} \tag{5.84}$$

The Dynamic Range is determined as

$$DR = \frac{P_S}{P_N} \tag{5.85}$$

It can be expressed as a function of the number of bits of resolution B. For a step size of Δ the maximum signal power for a sinusoidal signal is

$$P_S = \frac{1}{2}(2^B - 1)^2 \left(\frac{\Delta}{2}\right)^2 \tag{5.86}$$

and the noise is expressed as random with uniform amplitude distribution between $\pm\frac{\Delta}{2}$ as

$$P_N = \frac{1}{3}\left(\frac{\Delta}{2}\right)^2 \tag{5.87}$$

The dynamic range is then

$$DR^2 = \frac{3}{2}\left(2^B - 1\right)^2 \tag{5.88}$$

If the logarithm is taken of this last equation it reduces to the well known relationship between DR expressed in dB and the number of bits

$$DR_{dB} = 10\log(DR^2) = B \cdot 6\,dB + 1.76\,dB \tag{5.89}$$

B can be derived as

$$B = \frac{1}{2}\frac{\ln\left(\frac{2}{3}DR^2\right)}{\ln(2)} \tag{5.90}$$

If first the current I is solved by combining all of the above and then multiplied by the power supply voltage V_{DD}, an expression for the power is found as

$$P = N_N N_C kT\, V_{ov} V_{DD} \ln\left(\frac{2}{3}DR^2\right)\frac{DR^2}{P_S}BW \tag{5.91}$$

This expression shows the trade-off between power, dynamic range and bandwidth. It shows that increasing the DR or the BW requires more power. This is power necessary to settle to an accuracy of B bits for an input signal with power P_S, but only taking into account the effective g_m of the amplifier. So, only the current of the input stage is taken into account. Other effects play also a role, and are discussed later.

Let's give an example by applying expression (5.91) to the 900 mV ΔΣ modulator of Section 6.2. The factor N_N and N_C can be calculated as:

$$N_N \approx \left(1 + \frac{C_S}{C_{FB}}\right) + \left(\frac{C_{INT}}{\mathcal{F}_{dc} C_S}\right)^2 \left(\frac{2C_S}{C_{L,eff}}\gamma_w\right) \tag{5.92}$$

$$\approx (1+1) + \left(\frac{5}{(5/7)\,1}\right)^2 \left(\frac{2\,4}{16}\,13\right) = 20 \tag{5.93}$$

$$N_C = \frac{C_{L,eff}}{C_S} = 4 \tag{5.94}$$

Further are given: dynamic range of 77 dB, a BW of 16 kHz, V_{DD} of $0.9V$, V_{ov} is designed to be approximately $0.70V$ and $\gamma_w = 13$. Evaluating (5.91) with these values yields $P \approx 14\,\mu W$. This is about a factor of three less than the measured total power consumption. Expression (5.91) only takes into account the transconductance of the first integrator. There are, however, three integrators plus a comparator and the former have four current brances instead of two plus a common mode feedback amplifier. These contributions also need to be accounted for. From Table 6.3 we find that the first integrator draw 33 μA and its CMFB 40 μA. Further the 2nd and 3rd integrator each draw 6 μA and 9 μA for the CMFB. The comparator draws also 6 μA. All amplifiers exept the comparator have a duty cycle of 50% in the Switched Opamp Technique. So In total there is a good factor of 3 more current drawn than the bias current of the first OTA. We find a power consumption *estimate* of 49 μW. This is very close to the actual measured value of 40 μW.

In the introduction of this section the ambiguous effect of the power supply voltage was mentioned. Expression (5.91) can give information about it. It *looks like* the power is proportional to the V_{DD} and that reducing it cuts down on power consumption. This is unluckily only true as long as the input signal power can remain constant, which in general is not the case. It is clear from the expression that for low power consumption , the input signal amplitude (and thus its power) must be maximized. Hence, reducing the power supply in a realistic situation, brings about a smaller input range of the system, and thus a smaller applicable input signal and power. When lowering the power supply, also the maximum input amplitude lowers, but the amount depends on the particular circuit. It can always be assumed to be a proportion of the power supply $V_{S,max} = \rho_S V_{DD}$. When substituting this in (5.91) the power consumption has approximately the following dependency.

$$P \propto kTDR^2BW\frac{V_{ov}}{V_{DD}} \tag{5.95}$$

In the next section (5.91) and (5.95) are discussed as well as other power considerations not covered by these expressions.

5.5.2 *Discussion of Elements Influencing Power Consumption*

The final interest is the power consumption of the complete system, in this context a low voltage $\Delta\Sigma$ A/D converter. The total system power consumption depends on that of the separate building blocks, in casu the integrators, but also on the system level concept. Expression (5.95) can explain part of the circuit level issues, but low supply voltage gives rise to other sources of power increase, not taken into account by it. In the following the several elements influencing the power consumption to the worse or the better are discussed.

- The proportionality expression for the power consumption of an integrator (5.95) reveals that lowering the supply voltage *increases* the power consumption due to the fact that the input signal must scale down at the same time. This fact might be called the low voltage low power contradiction. Low voltage Switched Capacitor circuits are thus expected to show a less favorable power-resolution-bandwidth trade-off.

- The proportionality (5.95) also tells something about the region of operation of the transistors as well. For low power the over-drive voltage should be minimized. This is however only true as long as the condition for which the derivation is valid is not broken. It has been implicitly assumed that the capacitance excess factor N_C is not dependent on the over-drive voltage. When lowering V_{ov} for a constant current, the transistor width increased, and so does the C_{GS}. The latter is a parasitic load capacitance to the amplifier in the integrator that is seen in parallel with the sampling capacitor. In many low voltage applications such large C_S is required that for the required speeds of operation, the parasitic C_{GS} is negligible. Hence for low voltage and low power operation the over-drive voltage is best reduced as much as necessary, but not more. If the supply voltage is not too low a bias in the transition region between weak and strong inversion is recommended. The g_m/I ratio is almost as high as in deep weak inversion, while the parasitic capacitance is moderate.

- A complete amplifier has more branches of transistors than just the input transistors. The least possible number is in general an input and an output stage. For a differential amplifier this implies a minimum of four branches drawing current, as in a folded cascode or a symmetrical OTA. It has been shown in Chapter 4 that low voltage operation gives rise to extra branches. This is the case in the low voltage current mirror of Section 4.2 and also illustrated by the low voltage class AB OTA of Fig. 4.11. These current branches cause extra power consumption

A single stage amplifier causes less power consumption than a multi stage amplifier. Consider an integrator with a single stage amplifier to have a GBW of

$$GBW_{1s} = \frac{g_{m1s}}{C_{L,eff}} \tag{5.96}$$

$$C_{L,eff} = C_S + \frac{C_L}{\mathcal{F}_{dc}} \tag{5.97}$$

The index *1s* stands for 'single stage'. The GBW of an integrator with a two stage Miller compensated amplifier is given by

$$GBW_{2s} = \frac{\mathcal{F}_{dc}\, g_{m2s}}{C_C} \tag{5.98}$$

The index *2s* stands for '2 stage'. It has a non dominant pole of approximately

$$p_{nd} = \frac{g_{mout}}{\mathcal{F}_{dc} C_{L,eff}} \tag{5.99}$$

where g_{mout} is the transconductance of the amplifying transistor in the second stage. (The zero that exists is disregarded in this discussion, because it can be effectively compensated.) In this expression the output capacitance of the first stage is neglected for simplicity. The two structures must be compared for the same settling time, so for identical GBW, in the assumption that the integrator with single stage amplifier is critically damped by some non dominant pole, and the integrator with two stage amplifier by the p_{nd} at the output stage. Hence it is required to have

$$\frac{\mathcal{F}_{dc}\, g_{m2s}}{C_C} = \frac{g_{m1s}}{C_{L,eff}} \tag{5.100}$$

And for the system to be critically damped a ratio of 4 is required between the non dominant pole and the GBW,

$$\frac{g_{mout}}{\mathcal{F}_{dc}\, C_{L,eff}} = 4\,\frac{\mathcal{F}_{dc}\, g_{m2s}}{C_C} \tag{5.101}$$

The following relations can be derived:

$$g_{m2s} = g_{m1s}\frac{C_C}{\mathcal{F}_{dc}\, C_{L,eff}} \tag{5.102}$$

$$g_{mout} = 4\,\mathcal{F}_{dc}\, g_{m1s} \tag{5.103}$$

Assuming the same $V_{GS} - V_T$ between both cases if in strong inversion or assuming weak inversion the transconductances are proportional to the bias currents. Therefore the total current of the 2 stage opamp can be expressed as

$$I_{tot,2s} \quad \propto \quad I_{1s}\left(\frac{C_C}{\mathcal{F}_{dc} C_{L,eff}} + 4\mathcal{F}_{dc}\right) \tag{5.104}$$

where I_{1s} is the bias current associated with g_{m1s}. It is readily concluded that

$$I_{tot,2s} \quad > \quad 4\mathcal{F}_{dc}\, I_{1s} \tag{5.105}$$

A practical choice for C_C is about equal to $\mathcal{F}_{dc} C_{L,eff}$. Therefore the total current is approximately

$$I_{tot,2s} \quad \approx \quad (1 + 4\mathcal{F}_{dc})\, I_{1s} \tag{5.106}$$

A single stage amplifier has also an separate input and output stage, so the total current consumption of it is the double of the input stage. Hence the ratio of the current consumption between the two stage and single stage case is

$$r_I \approx \frac{(1+4\mathcal{F}_{dc})}{2} \tag{5.107}$$

which for a practical case is around a factor of two.

It can be concluded that due to the fact that a two stage amplifier is not load compensated, is has an increased power consumption of at least a factor of two for similar settling properties.

- The term N_C in (5.91) reveals that the effective load capacitance should be as small as possible. The expression for the Modified Switched Opamp integrator is given in (5.33) and (5.34) and repeated here for clarity:

$$C_{L,eff} = C_S + C_{CM} + C_{FB} + \frac{C_L + C_{CMS,eq}}{\mathcal{F}_{dc}} \tag{5.108}$$

$$\mathcal{F}_{dc} = \frac{C_{INT}}{C_S + C_{FM} + C_{CM} + C_{INT}} \tag{5.109}$$

  The presence of capacitors at the input of the integrator has a double disadvantage. They increase the effective load capacitance directly. Further they decrease the feedback factor. The total load capacitance at the output is divided by the feedback factor, so effectively it is made larger.

- The term N_N in (5.91) calls for the minimization of noise sources. Expressions (5.56) and (5.56) reveal that the noise is increased by the addition of capacitors at the input. There are several effects.

  The switch noise is directly increased. Especially C_{CM} in the Modified Switched Opamp integrator plays a large role, since it is only half the size of C_S and C_{FB}. For this case, the common mode adjust capacitor and the fact that the $V_{REF,\Delta\Sigma}$ must be sampled, increase the switch noise power by a factor of four.

  The opamp noise is increased through the feedback factor, which becomes small when more capacitors are added. For the opamp also the noise excess factors γ_w and γ_f are important. Coping with low supply voltage may cause large noise excess factor, as is the case for the low voltage class AB OTA and its CMFB amplifier of Section 4.4.6 and Section 4.5.5.

- In the analysis of previous section no CMFB circuitry is considered. The best option in conventional differential Switched Capacitor circuits is to use a Switched Capacitor CMFB. This uses the current source transistors in the output branches as feedback amplifier. Hence usually no additional current consumption is caused. However, in the Switched Opamp Technique it is not possible to apply the Switched Capacitor CMFB as explained in (3.4.4.1). The switches cannot be driven. Therefore an extra CMFB amplifier is required as shown in Section 3.4.4.2 or Section 3.4.4.3, or also by the practical realization of Fig. 4.15. This extra CMFB amplifier causes a significant amount of extra power consumption.

The settling speed of the CMFB, which is determined by its GBW and given by (4.105) requires attention. It is repeated here for clarity:

$$GBW = \frac{g_{meff,CM}}{2\pi C_{L,eff,CM}} \tag{5.110}$$

$$C_{L,eff,CM} = \frac{C_{CMS2}}{2} + \frac{C_L + C_{BP} + \frac{(C_S + C_{FB} + C_{CM})C_{INT}}{(C_S + C_{FB} + C_{CM} + C_{INT})}}{\mathcal{F}_{dc,CM}} \tag{5.111}$$

$$\mathcal{F}_{dc,CM} = \frac{C_{CMS}}{C_{CMS} + \frac{C_{CMS2}}{2}} = \alpha \tag{5.112}$$

The CMFB has to drive a large capacitive load. It sees the capacitances at the output of the integrator and also all of the capacitors at the input of the integrator in series with the integrating capacitor boosted by the feedback factor. Furthermore, in order to cope with the low supply voltage a special common mode sampling and dividing approach is followed (Section 4.5). For sub $1\,V$ implementations a realistic value is four, as applied in Section 6.2. The feedback factor of the CMFB system happens to be the division factor α, and thus the mentioned load capacitances are effectively boosted substantially, (i.e. by a factor of four). The necessity to apply the division on the common mode sample thus requires a larger CMFB amplifier transconductance in order to achieve a specific common mode settling accuracy. Hence a larger current needs to be drawn. Although the supply voltage is low, its beneficial effect on the power consumption is more than counterbalanced.

As a conclusion about the CMFB it can be said that the low supply voltage in combination of the Switched Opamp approach requires a dedicated CMFB amplifier. The latter one has a large capacitive load even further boosted by the necessity to divide the common mode sample to a low level where it can be applied at the input stage of an error amplifier with an input stage biased close to strong inversion.

- It is shown in Section 5.4.2 that the circuit noise generated inside the ΔΣ modulator loop is suppressed by a large factor. It is thus possible to allow larger noise sources inside the loop. The sampling capacitors can thus be chosen smaller. Because the corresponding opamps drive a smaller load they can be scaled down accordingly, which yields less power consumption. For low power operation it is thus imperative that the integrators are scaled down progressively.

  It was also shown in Section 5.4.3 that the quantization noise may not be performance limiting, because otherwise power is being wasted in driving an unnecessarily large capacitive load. Low power operation requires that the performance be circuit noise limited.

- There is basically only one option for power saving in a Switched Opamp integrator and it is on the system level. The intrinsic Switched Opamp integrator is a half delay integrator. It is off during 50% of the time. Since integrating and sampling of the output of an integrator happens simultaneously, a larger load than during each of the phases in the previous case needs to be driven. Because of the progressive down scaling of the integrators in a ΔΣ modulator, only a reduced bias current increase during the single on phase is required. This means that a power saving of almost 50% is possible. In the Original Switched Opamp Technique an extra half delay block is added, which requires an opamp and thus does not lower the power consumption at all. The system level concept must be adapted such that it allows to operate with half delay integrators, and without the addition of more integrators or delay elements. The major contribution described in this book towards low power is the new ΔΣ modulator topologies employing half delay integrators presented in Chapter 2.

The single loop topologies require no extra elements. The cascaded ones require one delay element per stage coupling.

From the above discussion the following low power guidelines are distilled, that cannot always be met for low voltage operation.

- Use the maximum input signal amplitude possible. In practice this implies to use the maximum supply voltage available.
- Bias the input transistors on the border between weak inversion and strong inversion.
- Reduce the number of current branches in circuits.
- Avoid introducing additional capacitors. They bring about increased load and noise sources.
- Try to use an amplifier with a low excess noise factor.
- Avoid to have to sample $V_{REF,\Delta\Sigma}$ if possible: use a double reference instead
- Avoid power hungry CMFB (error amp and division).
- Use a higher order $\Delta\Sigma$ modulator, and limit the performance by circuit noise.
- Implement a $\Delta\Sigma$ modulator with scaled down integrators in the loop.
- Use a rearranged system topology with half delay integrators. This allows to turn the amplifiers off during half of the time.

5.6 Conclusions

This chapter has treated design considerations in detail. First non-ideal charge transfer is studied. The transfer functions of the integrators are derived for a number of different OTA non-ideal properties. Behavioral simulations where these characteristics are varied allow to progressively determine the required specifications of the building blocks to ensure the overall system performance. The required specifications are quite relaxed as the single loop topology chosen for implementation. The optimum bias point of the OTA is investigated. A bias in weak inversion is more power efficient than a bias in strong inversion. Noise contributions in a $\Delta\Sigma$ modulator are studied, especially to investigate the influence of low voltage supply on it. Other influences on power consumption are studied. As a result design guidelines for low power are given. In general it can be concluded that low supply voltage increases the power consumption due to an intrinsic (physical) reason and practical reasons.

6 IMPLEMENTATIONS

In this chapter two implementations are presented. They are a $1.5V$- 100 μW $\Delta\Sigma$ modulator [Pel 97c] which is a single ended implementation of a second order modulator. The second implementation is a 900 mV - 40 μW $\Delta\Sigma$ modulator [Pel 98b][Pel 98a]. The latter is a fully differential implementation of a third order single loop topology.

6.1 A $1.5V$ - $100\mu W$ $\Delta\Sigma$ Modulator

The purpose of this design is to demonstrate the feasibility of low-voltage, low-power $\Delta\Sigma$ modulator ADCs based on the Switched Opamp Technique. The target specifications are set at 12 bit DR for the speech bandwidth.

6.1.1 The Architecture

For the intended performance, a second order single loop $\Delta\Sigma$ modulator sampled at 500 kHz offers enough quantization noise suppression and is therefore chosen for this design (see Table 2.2). A further benefit is that due to the inherent stability it is a good test vehicle for this feasibility study.

The block diagram of the second order $\Delta\Sigma$ modulator is shown in Fig. 6.1. For the integrator function $I(z)$ the normal full delay integrator function is implemented. The loop coefficients can be readily derived from Table 2.2. With the loop reorganized with all feedback coefficients unity, the optimal loop parameters are $(a_1, a_2) = (0.5, 0.5)$. In order to have the same signal distribution on the two integrator output nodes and have it equal to $V_{REF,\Delta\Sigma}$, the coefficients are scaled to the new set $(a_{1s}, b, a_{2s} = (1/3, 2/3, 1/2)$.

The Switched Opamp implementation of the topology of Fig. 6.1 is shown in Fig. 6.2. It consists of three OTAs. The second one performs the function of the delay element in a full delay

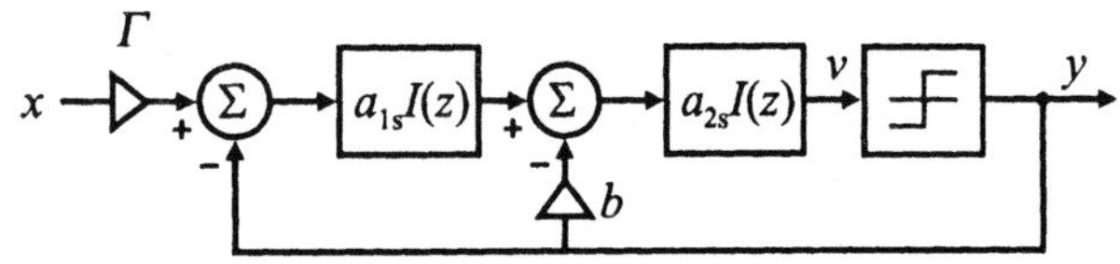

Figure 6.1. *ΔΣ modulator topology of 1.5V implementation.*

Switched Opamp integrator (Section 3.4.1.3, Fig. 3.11). The first and third OTA implement the integrating functions. The input switching scheme of Section 3.4.5.2 is implemented. It allows to maximally use the available input swing. In accordance to (3.45) and taking into account a stable input range of 0.75 and a desired input amplitude of 350 mV, the input scaling factor Γ is taken equal to 2. Only one feedback reference voltage $V_{REF,\Delta\Sigma}$ is used. Often, the feedback is realized with $+V_{REF,\Delta\Sigma}$ and $-V_{REF,\Delta\Sigma}$. This is, however, impossible with the Switched Opamp approach, because the level $+V_{REF,\Delta\Sigma}$, which is in the middle between the power supplies and does not allow a switch to be turned on, cannot be switched. Hence, only $-V_{REF,\Delta\Sigma}$ is used. The level to be switched is $V_{REF} - V_{REF,\Delta\Sigma}$, which is very close to V_{SS}, and thus allows for a very large over-drive of the switches connected to that voltage level. An inverting and non-inverting switching scheme are used to feed back with the correct sign, as can also be seen in Fig. 6.2. Two non-overlapping non-delayed clock phases are used. The Switched Opamp reference voltage V_{REF} is taken 500 mV. This leaves an over-drive voltage of 250 mV for the nMOST switches. The two stage Miller compensated single ended OTA of Fig. 4.1 has been used in this implementation.

6.1.2 Measurement Results

A test chip of the Switched Opamp ΔΣ modulator ADC is fabricated in a single-poly 0.7 μm CMOS process, with poly-diffusion capacitors. Fig. 6.3 shows a micro photograph of the test IC. Besides the ΔΣ it contains some overhead for testing purposes. The core area of the second order modulator takes up about 0.5 mm^2. The MOST devices in this technology have standard threshold voltages, being $V_{Tn} = 0.75V$ and $V_{Tp} = 0.95V$. The reference voltage can in practice be chosen somewhat higher than (3.44) and is $V_{REF,\Delta\Sigma} = 470\ mV$. The ΔΣ modulator is sampled at 500 kHz. The bit-stream output is processed off-line with a dedicated C-program, performing Blackman-Harris windowing [Har 78] , fast Fourier transform, and rectangular filtering of the signal band. The signal bandwidth complies with the speech bandwidth of 300-3400 Hz. Fig. 6.4 shows the *SNDR* versus the relative input amplitude $V_{IN}/V_{REF,\Delta\Sigma}$. The *DR* is 74 dB which is equivalent to 12 bits. The *PSNR* is 66 dB. The maximum input signal range is 120 mV_{rms}. Fig. 6.5 shows the measured spectrum in the signal band. The input signal corresponds to the *PSNR* and the signal frequency is 900 Hz, so up until the third harmonic are in the signal band. The harmonic distortion can be seen to be -72 dB. This is only *HD2*. The third harmonic is already negligible. This ΔΣ modulator ADC is operational down to a supply voltage of 1.2V. The full list of specifications is given in Table 6.1. [Pel 97b][Pel 97c]

6.2 A 900mV - 40μW ΔΣ Modulator

This design is an implementation of a ΔΣ modulator that has the lowest reported supply voltage hitherto and with the highest power-accuracy-speed figure of merit of all reported ΔΣ A/D converters with supply voltage lower than 2.5V. [Pel 98b][Pel 98a]

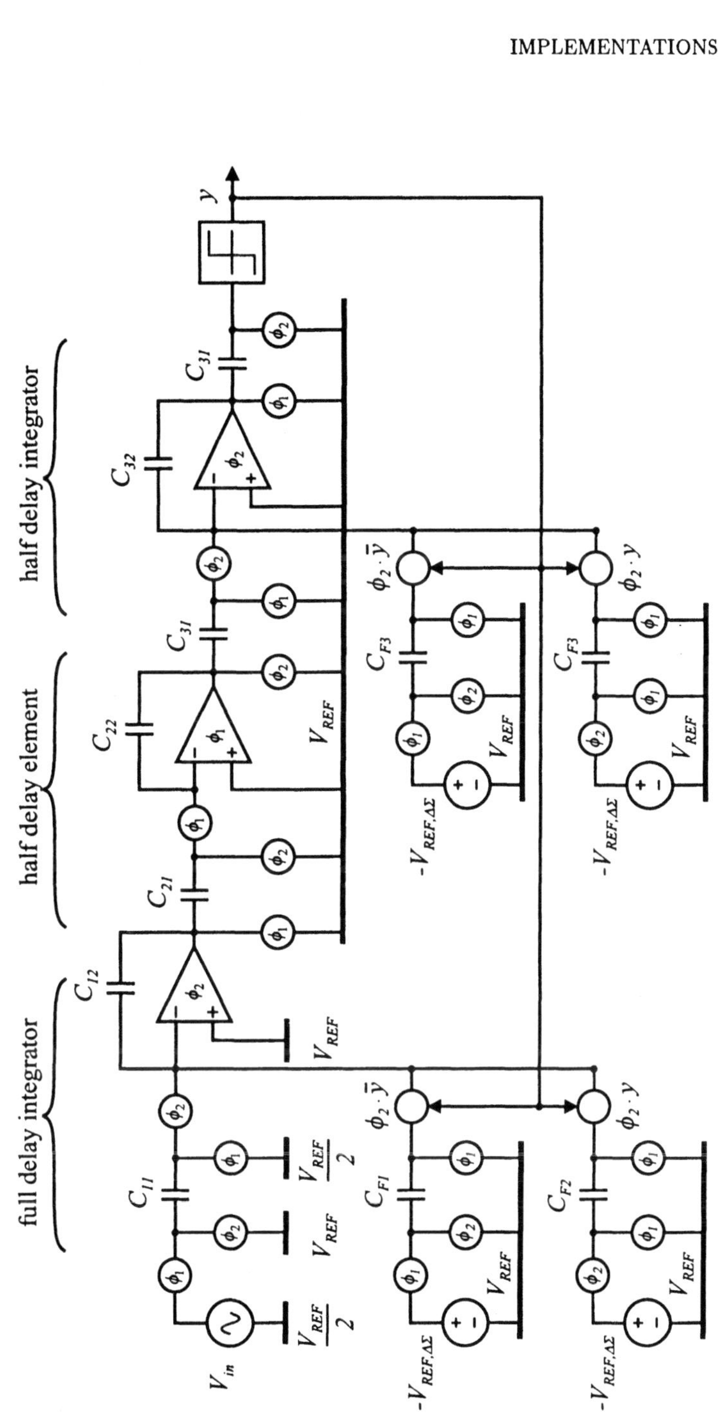

Figure 6.2. *Switched Opamp implementation of 1.5V $\Delta\Sigma$ modulator.*

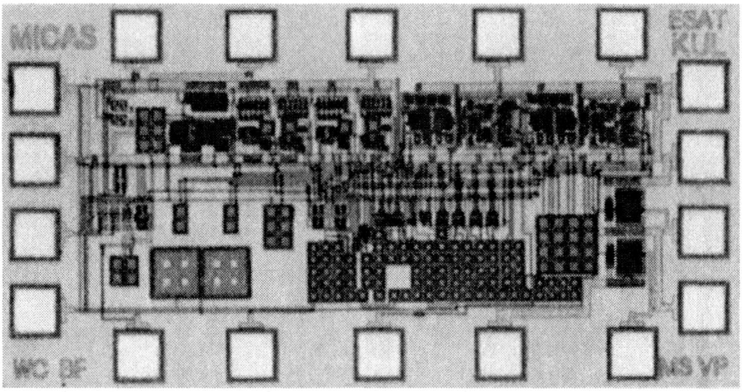

Figure 6.3. *IC micro photograph of the Switched Opamp implementation of the 1.5V ΔΣ modulator.*

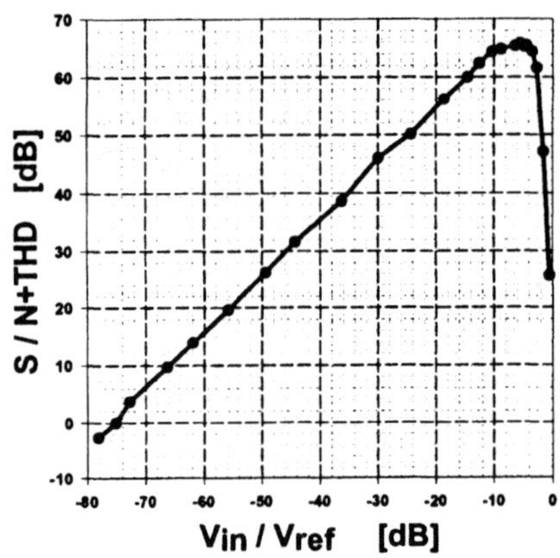

Figure 6.4. *The 1.5V ΔΣ modulator performance graph: the measured SNDR. The peak SNDR is 62 dB and the DR is 77 dB.*

6.2.1 Architecture

For this implementation a third order single loop topology is chosen. For low voltage and low power applications, a higher order single loop ΔΣ topology is preferable over a cascade topology because of its reduced sensitivity to linear integrator non-idealities, such as OTA gain, GBW and switch resistance. Low supply voltage makes it impossible to use cascode transistors in the output stage to increase the gain, while a two stage amplifier will consume for the same GBW a multiple of the power of a single stage one. A single loop architecture can operate with the gain provided by a one stage amplifier and therefore facilitates low power operation.

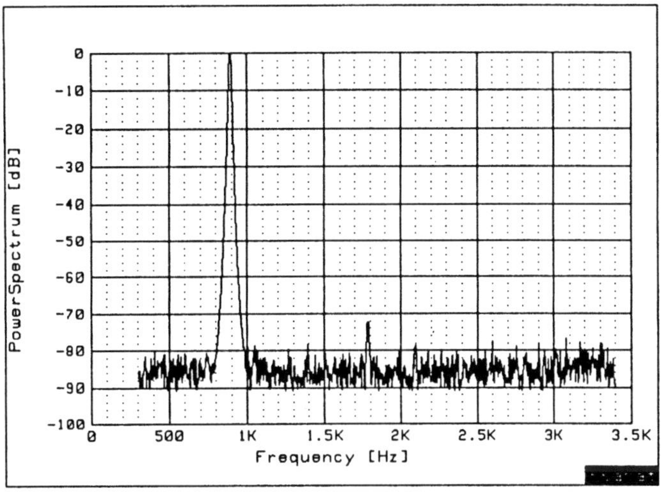

Figure 6.5. *The 1.5V ΔΣ modulator in-band spectrum for a full scale input signal. The second harmonic distortion is -72 dB. The third harmonic is negligible.*

Signal Bandwidth	300-3400	Hz
Sampling Frequency	500	kHz
max. input level	120	mV_{rms}
Dynamic Range	74	dB
peak *SNR*	66	dB
THD	< -72	dB
Power Supply	1.5	V
Power consumption	100	μW
chip core area	ca. 0.5	mm^2
Techology	0.7	μm CMOS

Table 6.1. *Performance summary of 1.5V ΔΣ modulator.*

In order to design for low power consumption it is important to remember the following from Section 5.4:

- The *SNR* must be limited by thermal noise (kT/C) and not by quantization noise.
- The first integrator needs a large sampling capacitor in order to reduce the kT/C noise to the required level
- Noise sources inside the loop are reduced, so the other integrators can be scaled down. In this design the sampling capacitors of the second and third OTA are 10 times smaller than the input sampling capacitor. Consequently the first integrator consumes the major part of the power.

The major power saving compared to the previous implementation of Section 6.1 is obtained on the architectural level by using the ΔΣ topology with half delay integrators of Fig. 2.4 rather

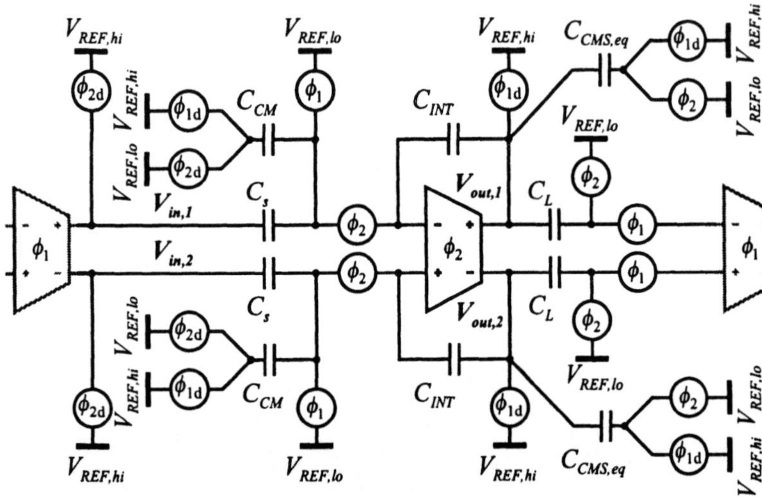

Figure 6.6. *The Differential Modified Switched Opamp integrator cell with delayed clocks.*

than the classic one of Fig. 2.3 with full delay integrators. As explained in Section 3.4.1.2 the Switched Opamp integrator is by nature a half delay integrator and a full delay integrator can only be obtained by adding delay blocks of analog nature that allocate an extra opamp (see Section 3.4.1.3 and Section 3.4.2.5). Hence they consume extra power. The half delay integrator $\Delta\Sigma$ topology described in this book (Section 2.3.1) allows to eliminate this power consumption and thus the overall power consumption is drastically reduced.

The $\Delta\Sigma$ modulator parameters are directly derived from Table 2.2. They are ($a_1 = 0.2, a_2 = 0.5, a_3 = 0.5$). The state variable distributions (v_1, v_2, v_3) are then bound by ($\pm 0.8 V_{REF,\Delta\Sigma}$, $\pm 1.25 V_{REF,\Delta\Sigma}$, $\pm 1.25 V_{REF,\Delta\Sigma}$). It is preferred not to scale the distributions to be equal. A smaller first state variable distribution is expected to yield a better linearity. Because the OTA gain varies less over a smaller output signal range, the static settling error is less signal dependent. This is even more important since the OTA is only a single stage amplifier and the DC gain is relatively limited (about 60 dB). With $V_{REF,\Delta\Sigma} = 200\,mV$ the signal excursions at the integrator outputs are 360 mV, 500 mV and 500 mV.

This implementation is fully differential. It makes use of the Differential Modified Switched Opamp integrator cell of Fig. 3.18, with delayed clocks as shown in Fig. 6.6. The first integrator cell is different because it must make use of the low voltage input sampling solution proposed in Section 3.4.5.2. The input signal is superimposed on a low common mode value of 100 mV. It is sampled with respect to that common mode value. Hence the sample capacitor doesn't have a DC component. The common mode adjustment capacitor C_{CM} is then not needed. The configuration of the first integrator in the system is shown in Fig. 6.7.

This integrator cell needs to be complemented with the feedback capacitors. The differential feedback network is shown separately in Fig. 6.8. Only one reference voltage is used. In the spirit of the Switched Opamp Technique only voltage levels in between the supply rails are allowed. This network is similar to the one shown in Fig. 6.2 in the sense that an inverting and non-inverting scheme are being used, but is different by the fact that $+V_{REF,\Delta\Sigma}$ is sampled and that it is differential. Where in the single ended feedback scheme one capacitor is idle during the integration, now both can be used each phase due to the differential nature.

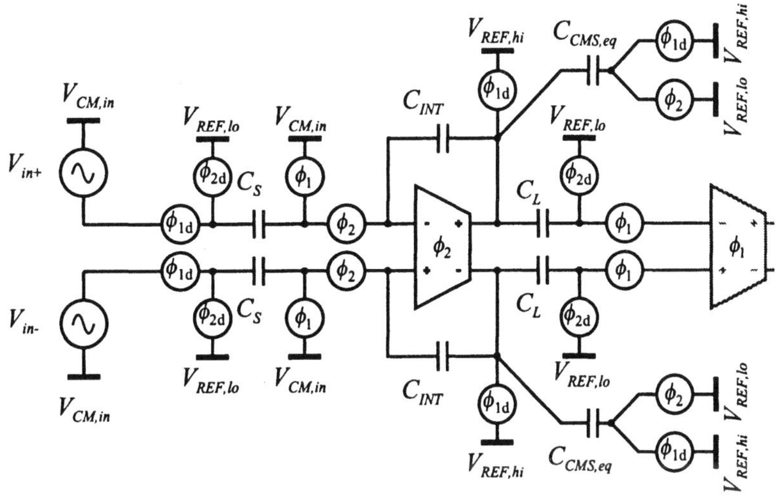

Figure 6.7. *The first integrator in the system features the low voltage sampling solution and hence does not have a C_{CM}. The input is applied symmetrically and sampled relative to a low common mode voltage. The maximum input level is low enough to be sampled by a switch.*

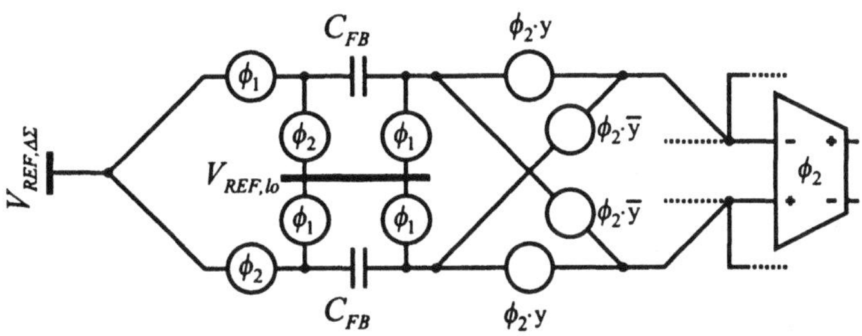

Figure 6.8. *The differential $\Delta\Sigma$ feedback scheme. The reference voltage is low enough to be switched. One reference is used and an inverting and non-inverting scheme are applied to do positive and negative feedback.*

The integrator cell makes use of the new low voltage class AB OTA of Section 4.4 and Fig. 4.11. The CMFB scheme is the one of Section 3.4.4.2 implemented as in Fig. 4.15. All transistors are biased in the moderate inversion region with $V_{GS} - V_T$ in the range of $-100\,mV$ to $+100\,mV$. All amplifiers are switched on and off. Table 6.2 lists all the capacitance values for the three integrators. $C_{Leff,OTA}$ and $C_{Leff,CMFB}$ are the effective load capacitances seen by the OTA and the CMFB amplifier. $C_{CMS,eq}$ is the equivalent capacitance of the common mode sampling network. The effective capacitances also take into account the main parasitic capacitances such as the bottom plate capacitance of the integrating capacitor. The input sampling capacitance is $4\ pF$ and the smallest unity capacitance used is $200\ fF$.

Fig. 6.9(a) shows the Bode plot for the first integrator. The integrators are designed to be critically damped. That this is so can be verified in Fig. 6.9(b) which is the transient simulation

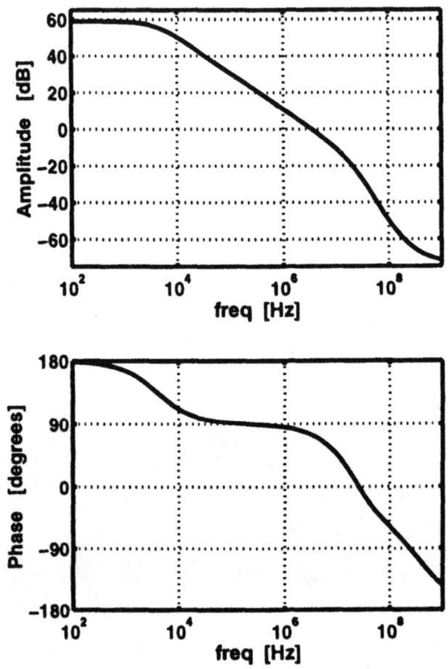

(a)

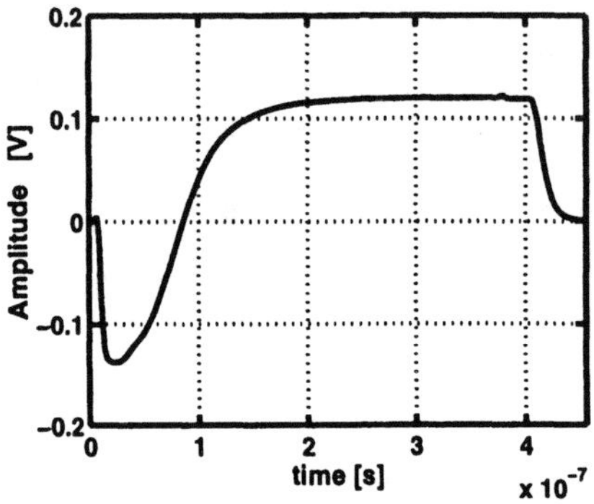

(b)

Figure 6.9. *Bode plot (a) and settling behavior (b) of the first integrator in the 900 mV ΔΣ modulator.*

OTA	C_S	C_{FB}	C_{CM}	C_{INT}	C_L	C_{CMS}	C_{CMS2}	$C_{CMS,eq}$	$C_{Leff,OTA}$	$C_{Leff,CMFB}$
1	4	4	0	20	0.4	2.66	16	2	14,2	42.1
2,3	0.4	0.4	0.2	0.8	0.4	0.66	4	0.5	3.25	6.2

Table 6.2. *The Capacitance values in pF for the 900 mV implementation. The effective load capacitances also take into account bottom plate capacitances.*

result for a full scale input signal with same sign feedback applied. All the designed OTA and CMFB amplifier specifications are shown in Table 6.3. Quite remarkable is the fact that the CMFB has such a small feedback factor, which makes the effective load capacitance so large that it behaves as a first order system, because the non-dominant poles are at a much higher frequency than the *GBW*. Consequently the power consumption of the CMFB amplifiers is 20% and 50% higher than the quiescent power consumption of the corresponding OTA.

Besides integrators some more building blocks are necessary to build a ΔΣ modulator. They are a comparator, a latch, some logic (nand) and a clock driver. The comparator developed is presented in Section 4.7.2. It has a total bias current of 6 μA. Building a latch and logic and a clock driver at 900 mV does not pose serious problems. The latch used here is shown in Fig. 6.10 [Yua 97]. The only precaution in designing the meta-stable point of the comparator output is that it must be chosen such that if the input of the comparator is below its resolution no improper trigger of the latch is induced. If the outputs of the comparators have not diverged

	OTA 1	CMFB 1	OTA 2,3	CMFB 2,3
Loop GBW (MHz)	4	2.2	3.9	2
$\mathcal{F}_{dc}$	0.714	0.25	0.444	0.25
LF gain (dB)	59	45	60	46
phase margin (°)	78	91	78	90
I_Q (μA)	33.2	40.1	6.24	9.36
settling error %	0.14		0.09	

Table 6.3. *OTA specs of 900 mV realization.*

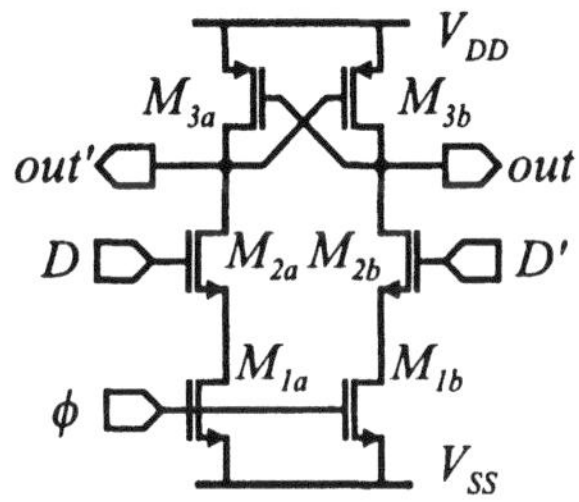

Figure 6.10. *The latch used in the 900 mV $\Delta\Sigma$ modulator.*

enough, in the available time the latch should not trigger. Hence the meta-stable point should be chosen below the threshold level of the latch, which here happens to be V_{Tn}.

6.2.2 Measurement Setup

The 900 mV $\Delta\Sigma$ modulator is implemented in a 0.5 μm standard CMOS n-well process with double poly and triple metal. The threshold voltages are $V_{Tn} = 620\,mV$ and $V_{Tp} = 550\,mV$. The IC micro photograph is shown in Fig. 6.11(a) The core area excluding bonding pads is 0.85 mm^2. Fig. 6.11(b) shows the functional blocks as long as the scale permits to distinguish them. From left to right the first, second and third integrators can be distinguished. The areas indicated by OTAi, include both the differential mode integrator OTA and the CMFB amplifier of which the layouts are interwoven. The CMFB fits in the middle between the two sides of the differential OTA and this compound layout is almost completely symmetrical, in spite of the non-differential nature of the CMFB amplifier. The switches turning on and off the OTAs are laid out all along the power rails over the entire width of the OTAs. In the first OTA the poly-poly capacitors are large enough to be distinguished: from left to right the CMFB capacitors, the integrating capacitors C_{INT} and the sampling C_S and feedback capacitors C_{FB} can be distinguished. Two U-shaped buses are located around the integrators. The inner bus is reserved for the analog interconnections, while the outer bus carries the digital signals: the clock and feedback lines. Given the very small currents, all interconnections are in principle minimal size. At the right of the digital bus the comparator and the latches are located. Above and below it sit clock drivers that together generate 8 clock signals from two external non-overlapping phases. All nMOST substrate contacts are connected to a single line, and have a separate bond-pad. This is contrary to the normal practice of precautions taken to counteract substrate noise, which require the bulk contacts of the devices in the bulk to be contacted on the spot to the source. The reason why this

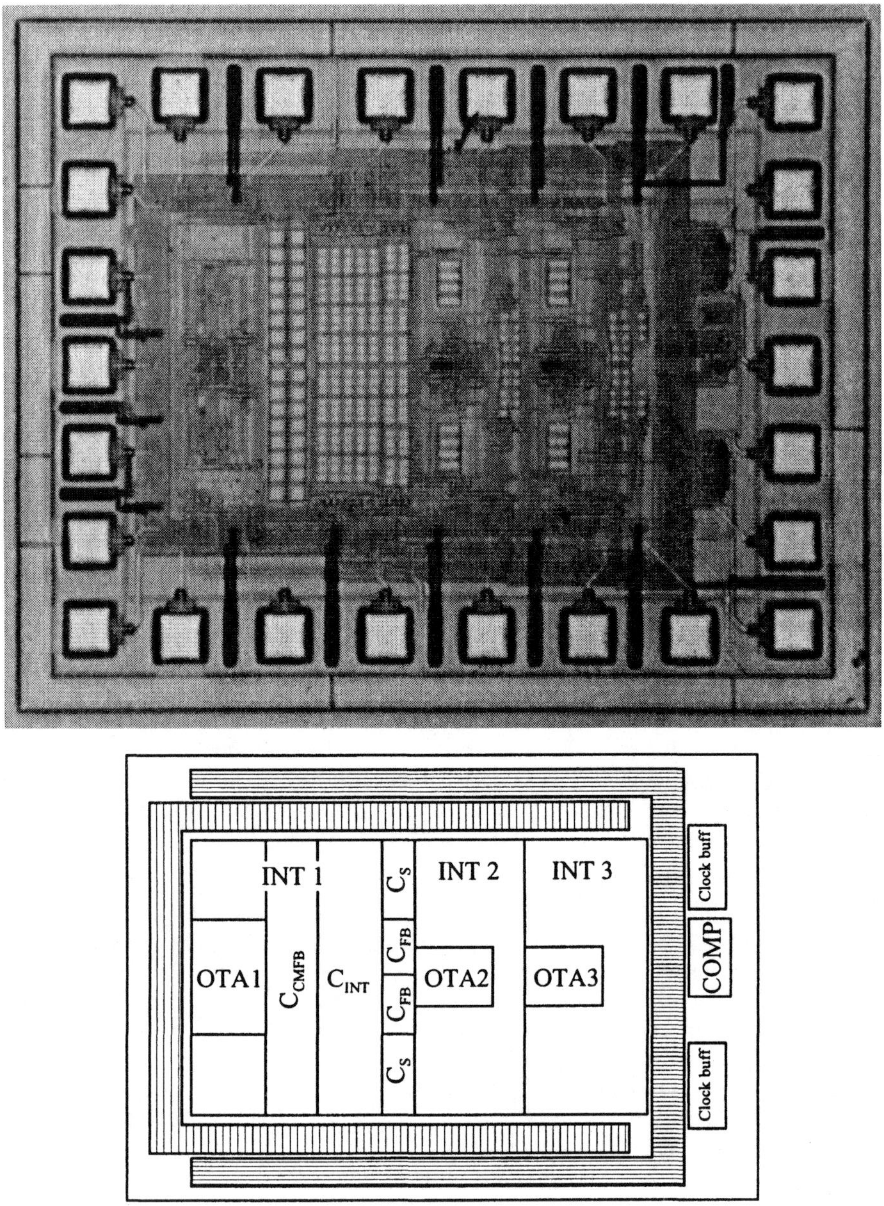

(a) IC floor plan.

Figure 6.11. *IC micro-photograph and floor plan. On the floor plan the three integrators with their OTA's and analog and digital interconnection busses are visible. The layout of the CMFB circuitry is interwoven with the OTA layout.*

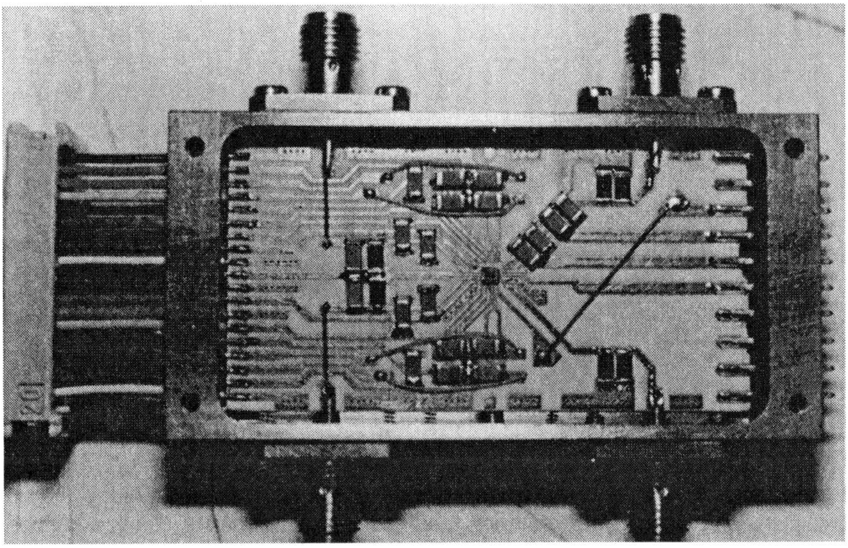

Figure 6.12. *The IC is mounted on a ceramic substrate and in a copper-berrylium box.*

is not done here is because the freedom to choose the bulk potential relative to the V_{SS} exists. In view of the feed forward signals below V_{SS} in the Modified Switched Opamp Technique, which actually forward biases slightly the junction diodes of the nMOST source diffusions of some switches, this possibility is built in to study the effect on the distortion. In practice the distortion did not vary whatsoever as a function of the bulk potential for negative and positive bias of a few hundred mV. Hence for the performance measurements the bulk is externally shorted to V_{SS}.

The IC is glued non-conductively on a thick film ceramic substrate and bonded with aluminum. The former is required to experiment with the IC substrate bias. The substrate holds decoupling capacitors. The ceramic substrate is mounted in a copper-beryllium box to shield the IC. A photograph of the opened box is shown in Fig. 6.12.

The complete measurement setup is depicted in Fig. 6.13. The IC is located in the center. The thick film substrate is represented by the rectangle around it. A number of signal lines are applied to the IC. The differential input signal is derived from a single ended signal generator through a transformer and applied relative to the common mode input level $V_{CM,in}$. An RC low pass filter with 20 kHz 3 dB frequency filters the input signal. The setup is powered with an external battery of $1.5V$. The IC supply voltage of 900 mV is derived with several resistive dividers. The analog, digital and buffer supplies (V_{DD_A}, V_{DD_D}, $V_{DD_{Buff}}$) each are derived separately. A 100Ω resistor in series with each of those allows to measure the drawn currents. That way the power consumption of the analog and digital parts can be measured separately. All bias voltages are also generated with resistive dividers. They are the $\Delta\Sigma$ reference voltage $V_{REF,\Delta\Sigma}$, the Modified Switched Opamp Technique reference voltages $V_{REF,lo}$ and $V_{REF,hi}$ (which in practice are set to V_{SS} and V_{DD}), the input common mode $V_{CM,in}$, common mode reference levels $V_{CM,1}$ and $V_{CM,2}$ (the former for the first integrator and the latter for the second and third) and the substrate bias V_{subs} which is actually derived from a separate battery to allow negative voltages, but which in practice is taken V_{SS}. The n-wells (V_{nwell}) are externally shorted to V_{DD}

Signal Bandwidth	200-16000	Hz
Sampling Frequency	1.538	MHz
max. diff. input level	500	mV_{pp}
DR	77	dB
peak *SNR*	76	dB
peak *SNDR*	62	dB
Supply voltage	900	mV
Power consumption	40	μW
chip core area	ca. 0.85	mm^2
Techology	0.5	μm std. CMOS

Table 6.4. *Performance summary of 900 mV $\Delta\Sigma$ modulator.*

by bond wires. The bias currents are generated with resistive dividers in series with the on-chip diode connected transistors biasing the current sources. The first OTA has its own current bias. The second and third OTA have a shared bias. Also a bias for the comparator current is present. The bias transistors are scaled up in order to allow reasonable currents of mA to flow externally. Hence they are not taken into account for the power calculation. Decoupling capacitors are connected between all bias points and V_{SS}. A pulse generator provides two non overlapping clock signals and a trigger for the logic analyzer that reads in the bit-stream output of the $\Delta\Sigma$ modulator. The logic analyzer is coupled to an HP workstation via the HPIB bus. On the workstation dedicated routines in Matlab perform FFT, windowing performance calculation and graphical representation of the measurements. Finally all cases of the equipment are shorted to V_{SS} and to earth.

6.2.3 Measurement Results

The measurement conditions are $V_{DD} = 900\,mV$, $V_{REF,\Delta\Sigma} = 200\,mV$, $V_{REF,lo} = V_{SS}$, $V_{REF,hi} = V_{DD}$, $V_{CM,in} = 100\,mV$, $V_{subs} = 0$. The bias currents are reduced to 75% of the designed values without loss of performance. The first integrator including CMFB consumes 60% of the total power. The second and third each consume 13% and the digital logic including clock drivers also consume 13%. The total core analog and digital power consumption is 40 μW. The sampling frequency is 1.538 MHz and the output bit-stream is processed with a comb filter for an oversampling ratio of 48.Fig. 6.14 shows the measured *SNR* and *SNDR*. The signal bandwidth is 16 kHz and the in-band noise is considered from 200 Hz on. The applied input signal frequency is 1.2 kHz. The measured peak *SNR* is 76 dB and the measured *DR* is 77 dB. The measured peak *SNDR* is 62 dB.

Fig. 6.15 shows two output signal spectra and the cumulative noise and distortion powers. Figure (a) is the output spectrum when no input signal is applied. The in-band noise floor is visible. Figure (b) is the output spectrum when the input signal corresonding to the *PSNR* is applied.

The measured performance is summarized in Table 6.4.

6.2.4 Conclusions

In this chapter two implementations of $\Delta\Sigma$ ADCs are presented. The system architecture, building blocks and measurement results are discussed. Both realizations are based on the

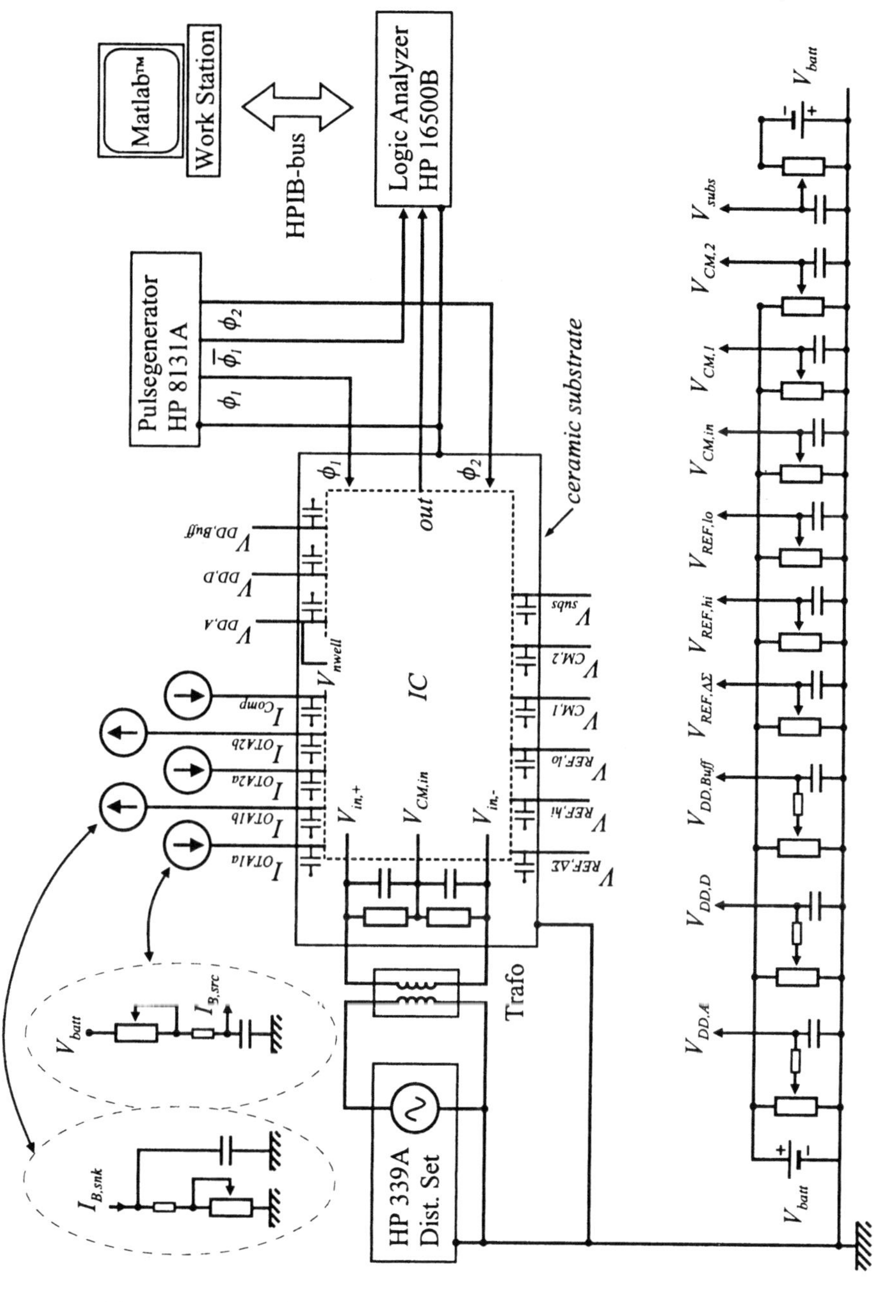

Figure 6.13. *The 900 mV ΔΣ modulator measurement setup.*

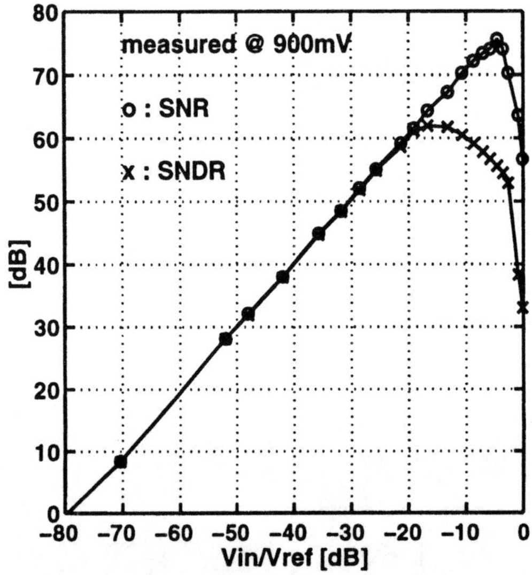

Figure 6.14. *The measured SNR and SNDR of the 900 mV ΔΣ modulator.*

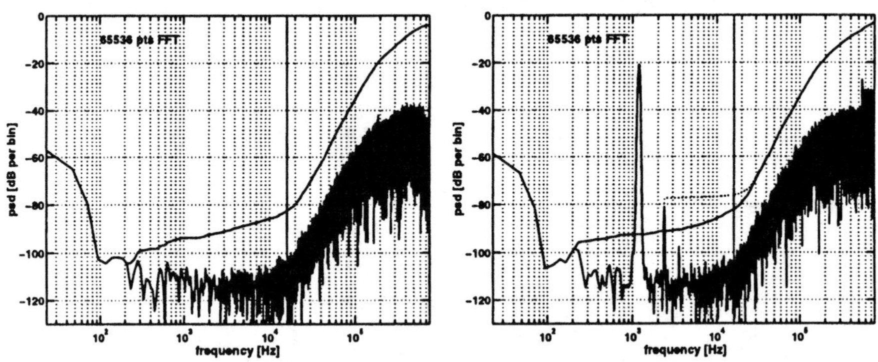

(a) *Noise floor* (b) *Spectrum with signal*

Figure 6.15. *The 900 mV ΔΣ modulator output spectrum. Also the cumulative noise and distortion powers are shown. (a) No signal applied, (b) Signal applied with amplitude corresponding to PSNDR*

Switched Opamp technique. The first one is an Original Switched Opamp implementation. This implies that it is a single ended implementation and that the achieved supply voltage is considerably high when compared to the next implementation. It also means that two opamps are needed for each integrator. It achieves 74 dB dynamic range in a signal bandwidth of 3.4 kHz and the power consumption is 100 μW. The THD is lower than 72 dB.

For the second implementation the Differential Modified Switched Opamp Technique is used. It thus allows lower supply voltage operation of 900 mV and it is a differential implementation. Also only one opamp per integrator is needed. Using a modified single loop $\Delta\Sigma$ modulator topology that only needs half delay integrators, an important part of the power consumption is avoided compared to the previous implementation. The measured performance is 77 dB DR in 16 kHz signal bandwidth and the measured power consumption is 40 μW. The peak $SNDR$ is 62 dB.

The performance of these $\Delta\Sigma$ ADCs can be compared to other low voltage implementations of Table 1.1 in Chapter 1. The 1.5V implementation has a good power-resolution-bandwidth figure of merit. It must be realized that compared to a Switched Capacitor implementation with voltage multipliers or low V_T transistors or a continuous time implementation with higher supply voltage, the input range of a Switched Opamp implementation is much reduced, which is very disadvantageous for power consumption. The 900 mV implementation has the highest figure of merit, even though it has the lowest supply voltage of all.

7
FINAL DISCUSSION

7.1 Conclusions

In this book the full design path of low-voltage low-power CMOS $\Delta\Sigma$ A/D converters has been treated, including system level, implementation level, circuit design techniques, power and design considerations as well as several realized IC implementations.

In Chapter 2 the classic single loop and cascade $\Delta\Sigma$ modulator topologies are modified to operate with half delay integrators. This is done because the basic Switched Opamp integrator cell constitutes a half delay. An additional delay element requiring an amplifier can thus be avoided. This measure causes a considerable power saving.

A systematic study of the mentioned classic $\Delta\Sigma$ modulator topologies is described based on [Pel 97a] [Mar 97]. It has allowed to find the optimal coefficients for those topologies with respect to peak *SNR*. For the single loop topology, the results are obtained through ample behavioral simulations, while for the cascade a design procedure provides guidelines for the choice of the parameters. The performance degradation is quantified by means of a linear model.

In Chapter 3 the key problem of low voltage Switched Capacitor signal processing and the solutions are discussed: it is the switch driving problem. A detailed exposition of the status of the Switched Opamp Technique is given. The Original Switched Opamp Technique , Modified Switched Opamp Technique and Differential Modified Switched Opamp Technique are discussed. Existing CMFB techniques are treated. A new CMFB sampling configuration is presented. The Switched Opamp Technique allows to operate Switched Capacitor integrators with only the low supply voltage, by eliminating the sampling switches that are not connected to a fixed voltage. A sampling switch at the input of the system can, however, not be avoided and a limited input range results. A new input sampling configuration making maximally use of the available input range is developed.

In Chapter 4 the influence of the low supply voltage is studied. At very low voltage, ensuring enough output signal swing becomes a greater limitation that the threshold voltage to further reduce the supply voltage. After the discussion of a two current feedback schemes that can be used as a low voltage current mirror, a new input stage is presented which consists of an addition of a transistor to the mentioned structure. These two building blocks are the cornerstones of a low voltage class AB OTA that is particularly suited for Differential Modified Switched Opamp applications. A dedicated CMFB amplifier is presented based on a 'triple differential' or hence called 'trifferential' input stage. A full analysis including pole-zero analysis, noise analysis and a design procedure is given. A new way of switching the amplifiers on and off is presented: putting switches between the internal and external power rails allows to achieve higher switching speeds. Other required building blocks such as the comparator are discussed and a new low voltage regenerative comparator is presented.

In Chapter 5 practical design considerations are discussed. From modeling the non-ideal characteristics of the opamps in a Switched Opamp integrator, and performing behavioral simulations with those, the required opamp specifications are derived. Designing the circuits with the found specifications ensures that the required performance is met. The influence of the region of operation of the MOSTs in an input stage is studied. Operation in weak inversion causes less power consumption. It is proven that for low power consumption a $\Delta\Sigma$ ADC must be designed such that its performance is limited by circuit noise and not by quantization noise. The derivation of an expression for the power consumption of any Switched Capacitor integrator is presented. It reveals many influences on the power consumption, which are discussed. As a general conclusion it can be said that low voltage operation and low power operation are contradictory requirements for a Switched Capacitor circuit.

In Chapter 6 two implementations are discussed. The first [Pel 97c] is a single ended $\Delta\Sigma$ modulator based on the Original Switched Opamp Technique operating with $1.5V$ supply voltage and consumes $100\ \mu W$ for a dynamic range of $74\ dB$ in a signal bandwidth of $3.4\ kHz$. This was the first $\Delta\Sigma$ ADC implemented with the Switched Opamp Technique and proved that the Switched Opamp Technique can be successfully applied to the design of $\Delta\Sigma$ ADCs. The second implementation [Pel 98b] is a fully differential implementation based on the Differential Modified Switched Opamp Technique. It consumes $900\ mV$ supply voltage and $40\ \mu W$ for a dynamic range of $77\ dB$ in $16\ kHz$ bandwidth. The latter implementation is besides the lowest supply voltage $\Delta\Sigma$ ADC also the lowest supply voltage Switched Capacitor circuit. Nonetheless it has an excellent power-resolution-bandwidth figure of merit. Hence it is proven that high quality signal processing is possible with the Switched Opamp technique and Switched Capacitor techniques in general at low supply voltage.

7.2 Possibilities for Future Work

The implementations in this book show that the Switched Opamp Technique allows to design high quality $\Delta\Sigma$ ADCs at very low supply voltage. This can be generalized to say that high quality Switched Capacitor operation at low voltage is possible. The driving force behind the $900\ mV$ implementation has been to achieve sub $1\ V$ supply voltage operation for a signal bandwidth of the order of $10\ kHz$. The possibilities for future work are twofold. Now that the sub one volt operation has been achieved featuring the considerably high dynamic range of $77\ dB$ in a $16\ kHz$ bandwidth, the immediate following challenge is to increase the performance to audio specifications, albeit for a relaxed supply voltage specification of $1.2V$. This means a dynamic range corresponding to 16 bits ($98\ dB$) in a $20\ kHz$ signal bandwidth. The question is whether the Switched Opamp Technique would at this point be the implementation method of choice. The main driving source for the development of the Switched Opamp Technique is that in (future) deep sub-micron processes voltage multipliers cannot be applied, because the MOST gate oxides will not sustain the multiplied voltage. For the main reason that a Modified Switched Opamp integrator cell requires a number of capacitors and that a Switched Capacitor CMFB cannot be applied, the Switched Opamp Technique is not so favorable for low power operation. A positive aspect for power consumption is the fact that the amplifiers are off during half of the time. Another solution towards the switch problem would be to not apply voltage multiplication to drive the switches, but some sort of a bootstrap that puts the maximum allowed voltage over the gate source terminals. Recently such a bootstrap circuit has been developed at the University of California at Berkeley [Abo 96]. The best approach for a low power audio $\Delta\Sigma$ ADC at $1.2V$ would be to apply the best of two worlds. This means the use of a half delay integrator cell with switched opamps and the decoupling of the DC levels at the input and output as in the Modified Switched Opamp Technique. For the switches the mentioned bootstrap can be applied. At the same time the input sampling problem is solved, which allows to apply the normal (large) $1V$ audio input signal, and thus has a tremendous beneficial impact on power consumption. It has the further benefit that it keeps the switch resistance more constant and thus linear. Another approach would be to tackle the input problem, which is being investigated [Bas 98]. Since the first integrator is not preceded by an OTA, the Switched Opamp Technique cannot be applied there. Preceding it with a switched buffer would be a solution. A new type of switched buffer uses resistive feedback and a Switched Capacitor generated DC level shift between the input DC level and the input pin of the OTA which sits at V_{SS} level.

The second challenge concerns pushing the speed of operation. After achieving the high resolution, the objective is to design low voltage low power high resolution $\Delta\Sigma$ ADCs and to find out if and when turning on and off the amplifiers becomes a performance limiting factor.

Appendix A
Calculations for Chapter 3

A.1 Efficiency Calculation of the Dickson Multiplier

In this appendix the efficiency of the Dickson multiplier is calculated. Efficiency is defined as

$$\eta = \frac{P_U}{P_D} = \frac{P_U}{P_U + P_L} = \frac{1}{1 + \frac{P_L}{P_U}} \quad \text{,where} \tag{A.1}$$

$$P_U = \text{useful power} \tag{A.2}$$

$$P_S = \text{power delivered by sources} \tag{A.3}$$

$$P_L = \text{lost power} \tag{A.4}$$

In order to derive an expression for the efficiency, it suffices to write down expressions for the useful and lost power terms. In this analysis, we'll look at the physical processes in one phase. Since power is energy by time, it's more convenient to talk about energy. At the end the conversion to power is easily made. In a charge pumping circuit there are typically two types of energy losses, based on the same reason. Charge redistribution between two capacitors always involves loss of energy. Let's illustrate this and at the same time calculate the energy loss caused by the presence of a stray capacitor.

Consider one capacitor C and its stray capacitor C_s out of the Dickson chain of Fig. 3.5. It is shown in Fig. A.1. Their top plates are always connected. Suppose they are both charged to V_I and that in the next phase, the capacitor C is lifted by the source V_ϕ. This causes a charge

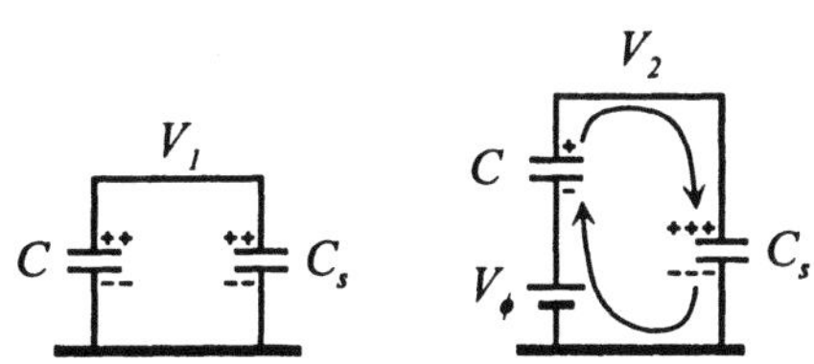

Figure A.1. *Charge redistribution between a capacitor and its stray capacitor in a voltage multiplier chain.*

redistribution between C and C_s. The new voltage is

$$V_2 = V_I + V'_\phi \qquad \text{where} \qquad V'_\phi = \frac{C}{C + C_s} V_\phi \tag{A.5}$$

(For V_ϕ see (3.5)) In order for the redistribution to take place, charge needs to flow, as indicated by the arrows. Let's go back to the original definition of a voltage source. It is a seat of *Electromotive Force*. It is capable of applying work to a charge in order to bring it up to a higher potential. For the redistribution current to flow, that is exactly what it needs to do. The *EMF* $\mathcal{E}$ is defined as

$$\mathcal{E} = dW/dq, \tag{A.6}$$

of which the units are *Volt* = *joule/coulomb*. Therefore the total work done on a charge is

$$W = \mathcal{E} q = V_\phi q \tag{A.7}$$

Because a voltage source is assumed here to have zero internal resistance, its potential V_ϕ equals its *EMF* $\mathcal{E}$. An additional amount of charge equal to

$$q = C_s V'_\phi \tag{A.8}$$

has been stored on C_s (while the same amount has disappeared from C). The work done is thus

$$W = C_s V'_\phi V_\phi \tag{A.9}$$

Now consider the next phase where the two capacitors are returning to their original state shown on the left. The charge redistributes again to its original situation. Between two successive identical states, there is conservation of charge and the total charge is the same. Therefore, both capacitors store the same energy as before, although in the intermediate state, energy has been pumped into the circuit equal to (A.9). This energy has not been delivered to a load as useful energy, but it has been dissipated as heat in the interconnections. The work necessary for the redistribution charge to flow has been lost. So the presence of a stray capacitor causes energy loss $E_{L,s}$ equal to W:

$$E_{L,s} = C_s V'_\phi V_\phi \tag{A.10}$$

This happens once per period so the lost power is

$$P_{L,s} = C_s V'_\phi V_\phi f \tag{A.11}$$

For N stages, the total lost power is

$$P_{L,s} = N C_s V'_\phi V_\phi f = N C_s \frac{C}{C + C_s} V_\phi^2 f \tag{A.12}$$

Now we'll proceed to calculate the second type of energy loss related to the use of capacitors. Consider one charge pumping stage out of the diode chain, without stray capacitors as shown in Fig. A.2. If the load at the end of the chain is drawing a current I, then the voltage in the last capacitor droops. The charge loss in one phase is

$$q = C \Delta V_{out} = I/f \tag{A.13}$$

This charge loss is replenished through the chain, and this action causes the voltage drop to ripple through the entire chain. So each capacitor sees a voltage drop corresponding to the

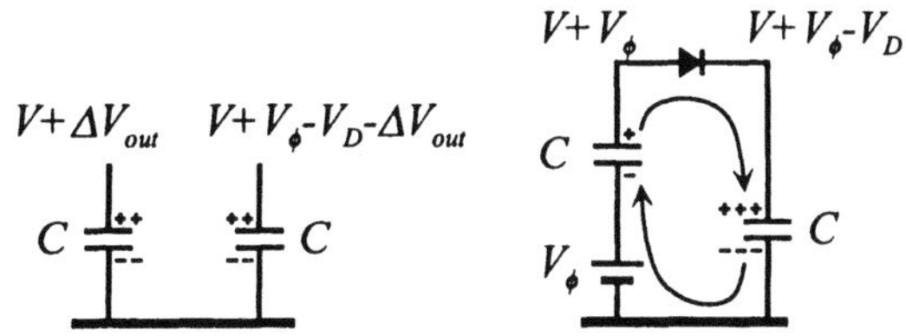

Figure A.2. *Charge redistribution between two consecutive capacitors in a voltage multiplier chain.*

charge drawn by the load in one phase. On the left of the figure the left capacitor has just been replenished to $V + \Delta V_{out}$, while the one on the right has just delivered an amount of charge to the next stage, corresponding to a voltage drop of ΔV_{out}. The difference in energy stored on the capacitors can be calculated to be

$$\Delta E = -C(V_\phi - V_D - \Delta V_{out})\Delta V_{out} \tag{A.14}$$

The total energy balance can be written:

$$C\Delta V_{out} V_\phi = C(V_\phi - V_D - \Delta V_{out})\Delta V_{out} + C V_D \Delta V_{out} + C\Delta V_{out}^2 \tag{A.15}$$

The left hand side is the work the source needs to do on the redistribution current. The first term of the right hand side is the useful energy increase, stored on C. The middle term is the energy loss the transferred charge undergoes by the voltage drop in the diode. And the last term is the energy dissipation caused by the charge redistribution.

Let's look at this last term and bring it in relation with the internal resistance. The total voltage drop in N stages is

$$\Delta V_{out,N} = \frac{I}{N f C} \tag{A.16}$$

From this expression the concept of equivalent internal resistance can be defined. The current drawn causes an internal voltage drop. The equivalent internal resistance is given by

$$R_{eq} = \frac{1}{N f C} \tag{A.17}$$

This equivalent internal resistance dissipates power as if it were a real resistor. The physical reason for it, however is charge redistribution. In fact a more elaborate calculation would reveal that [Dic 76]

$$R_{eq} = \frac{1}{N f (C + C_s)} \tag{A.18}$$

The third case of energy dissipation appeared already in the energy balance of (A.15) and is caused by the voltage drop over the diode. The charge flowing through the diode loses potential, which is a pure loss. It is given by

$$\Delta E = C\, \Delta V_{out} V_D \tag{A.19}$$

or when taking into account also the parasitic capacitor

$$\Delta E = (C + C_s)\, \Delta V_{out} V_D \tag{A.20}$$

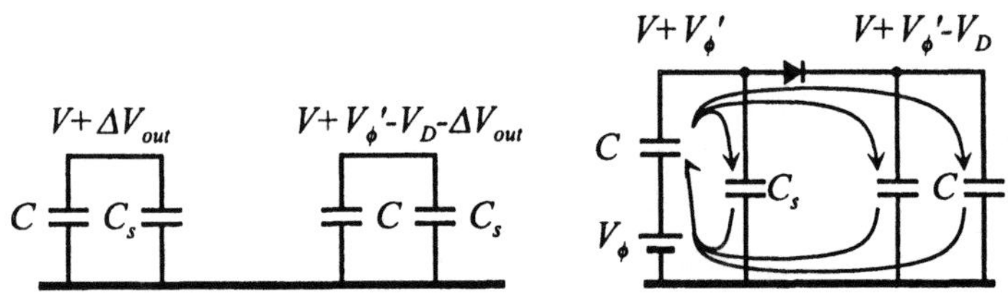

Figure A.3. *Charge redistribution between consecutive capacitors and their stray capacitors in a voltage multiplier chain*

For an N stage multiplier chain we can find the lost powers

$$P_{L,R_{eq}} = Nf\,(C + C_s)\Delta V_{out}^2 \tag{A.21}$$

and

$$P_{L,D} = Nf\,(C + C_s)\Delta V_{out} V_D \tag{A.22}$$

We could also write the energy balance for one pumping stage out of a chain, including the stray capacitors. Looking at Fig. A.3 it can be visualized that the source needs to do the work on the charge flowing into C_s on the left and separately the work on the charge flowing into $C + C_s$

$$E = C_s V_\phi V_\phi' + (C + C_s)\Delta V_{out} V_\phi \tag{A.23}$$

As explained above, the first term in (A.23) will be completely lost while the second term including the stray capacitors now equates

$$(C + C_s)(V_\phi - V_D - \Delta V_{out})\Delta V_{out} + (C + C_s)V_D\Delta V_{out} + (C + C_s)\Delta V_{out}^2 \tag{A.24}$$

The useful power delivered to a load drawing a current I is

$$P_U = I\, V_{out} \tag{A.25}$$

Now all the elements are given to calculate the efficiency. In fact not the efficiency is given here but the ratio of lost and useful powers.

$$\frac{P_L}{P_U} = \left(\frac{N}{V_{out}}\left(\frac{\alpha}{(1+\alpha)}\,V_\phi^2\,\frac{C\,f}{I_{out}} + (1+\alpha)\frac{I_{out}}{C\,f}\right) + (1+\alpha)\frac{N V_D}{V_{out}}\right) \tag{A.26}$$

where

$$\alpha = \frac{C_s}{C}. \tag{A.27}$$

A.2 Calculation of the required C_{CM}

The calculation of the required C_{CM} of the Modified Switched Opamp integrator of Fig. 3.14 or Fig. 3.16 is carried out here. During ϕ_I the charges on the capacitors are:

$$\begin{cases} q_{S_{n,I}} &= C_S(V_{DC@OUT} - V_{REF,lo} + V_S) \\ q_{CM_{n,I}} &= 0 \\ q_{INT_{n,I}} &= C_{INT}\,V_{OUT_{n-1,2}} \end{cases} \tag{A.28}$$

During ϕ_2 the charges on the capacitors are:

$$\begin{cases} q_{S_{n,2}} & = & 0 \\ q_{CM_{n,2}} & = & C_{CM}(V_{REF,hi} - V_{REF,lo}) \\ q_{INT_{n,2}} & = & C_{INT}V_{OUT_{n,2}} \end{cases} \tag{A.29}$$

The following holds

$$V_{DC@OUT} = \frac{V_{REF,lo} + V_{REF,hi}}{2} \tag{A.30}$$

The equations for conservation of charge are:

$$\begin{cases} q_{S_{n,1}} + q_{CM_{n,1}} + q_{INT_{n,1}} = q_{S_{n,2}} + q_{CM_{n,2}} + q_{INT_{n,2}} \\ q_{INT_{n,1}} = q_{INT_{n-1,2}} \end{cases} \tag{A.31}$$

This yields

$$C_S(V_{DC@OUT} + V_S) + C_{INT}V_{OUT_{n-1,2}}$$
$$= C_{CM}(V_{REF,hi} - V_{REF,lo}) + C[INT]V_{OUT_{n,2}} \tag{A.32}$$
$$V_{OUT_{n,2}} = V_{OUT_{n-1,2}} + \frac{1}{C_{INT}}(C_S V_S + C_S\frac{V_{REF,hi} - V_{REF,o}}{2}$$
$$- C_{CM}(V_{REF,hi} - V_{REF,lo})) \tag{A.33}$$

For proper integration, the two last offset terms must vanish. This is accomplished for

$$C_{CM} = \frac{C_S}{2} \tag{A.34}$$

A.3 Calculation of the required $C_{CM,aux}$

Here the required $C_{CM,aux}$ needed in the Modified Switched Opamp integrator of 3.15 is calculated. During ϕ_1 the charges on the capacitors are:

$$\begin{cases} q_{S_{n,1}} & = & C_S(V_{DC@OUT} - V_{REF,lo} + V_S) \\ q_{CM_{n,1}} & = & 0 \\ q_{INT_{n,1}} & = & C_{INT}(V_{REF,lo} - V_0) \\ q_{CM,aux_{n,1}} & = & C_{CM,aux}(V_{REF,hi} - V_0) \end{cases} \tag{A.35}$$

During ϕ_2 they are:

$$\begin{cases} q_{S_{n,2}} & = & 0 \\ q_{CM_{n,2}} & = & C_{CM}(V_{REF,hi} - V_{REF,lo}) \\ q_{INT_{n,2}} & = & C_{INT}V_{OUT_{n,2}} \\ q_{CM,aux_{n,2}} & = & 0 \end{cases} \tag{A.36}$$

The charge conservation equations are:

$$\begin{cases} q_{S_{n,1}} + q_{CM_{n,1}} + q_{INT_{n,1}} + q_{CM,aux_{n,1}} = q_{S_{n,2}} + q_{CM_{n,2}} \\ \qquad + q_{INT_{n,2}} + q_{CM,aux_{n,2}} \\ q_{INT_{n-1,2}} + q_{CM,aux_{n-1,2}} = q_{INT_{n,1}} + q_{CM,aux_{n,1}} \end{cases} \tag{A.37}$$

Solving these equations taking also into account that $C_{CM} = C_S/2$ one finds

$$V_{OUT_{n,2}} = V_{OUT_{n-1,2}} + \frac{C_S}{C_{INT}}V_S \tag{A.38}$$

This means that the integrator transfer function is guaranteed without condition for $C_{CM,aux}$. This is plausible because this capacitor only serves as intermediate storage of the integrator charge.

The condition for $C_{CM,aux}$ is to be found from a requirement for the minimum voltage V_O at the inverting Opamp input during the off phase $\phi 1$. It is desired that

$$V_O \geq 0 \tag{A.39}$$

From the second charge equation of (A.37) an expression for V_O can be found.

$$\begin{aligned} V_O &= -\frac{C_{INT}}{C_{INT}+C_{CM,aux}} V_{OUT_{n-1,2}} + \frac{C_{INT}}{C_{INT}+C_{CM,aux}} V_{REF,lo} \\ &\quad + \frac{C_{CM,aux}}{C_{INT}+C_{CM,aux}} V_{REF,hi} \end{aligned} \tag{A.40}$$

The two worst case conditions, which are in reality never met, because the output swing is not completely rail-to-rail, are

$$\begin{cases} a) \quad V_{OUT_{n-1,2}} = V_{REF,lo} \quad : \quad V_O = \dfrac{C_{CM,aux}}{C_{INT}+C_{CM,aux}} V_{REF,hi} \\ b) \quad V_{OUT_{n-1,2}} = V_{REF,hi} \quad : \quad V_O = \dfrac{C_{CM,aux}-C_{INT}}{C_{CM,aux}+C_{INT}} V_{REF,hi} \\ \qquad\qquad\qquad\qquad\qquad\qquad + \dfrac{C_{INT}}{C_{INT}+C_{CM,aux}} V_{REF,lo} \end{cases} \tag{A.41}$$

Expression a) is always positive, so the condition is to be found from b)

$$V_O \geq 0 \quad \Leftrightarrow \quad C_{CM,aux} \geq C_{INT}\left(1 - \frac{V_{REF,lo}}{V_{REF,hi}}\right) \tag{A.42}$$

A sufficient and at the same time practical condition is

$$C_{CM,aux} = C_{INT} \tag{A.43}$$

In this case the two extreme intermediate voltages are

$$\begin{cases} a) \quad V_{OUT_{n-1,2}} = V_{REF,lo} \quad : \quad V_O = \dfrac{V_{REF,hi}}{2} \\ b) \quad V_{OUT_{n-1,2}} = V_{REF,hi} \quad : \quad V_O = \dfrac{V_{REF,lo}}{2} \end{cases} \tag{A.44}$$

It can be seen that V_O is a compressed and inverted version of V_{OUT}.

A.4 Calculation of the Modified Switched Opamp Integrator Transfer Function

Here it is proven that the corrected Modified Switched Opamp integrator cell of Fig. 3.16 has a normal integrator transfer function. During ϕ_1 the charges can be expressed as:

$$\begin{cases} q_{S_{n,1}} &= C_S(V_{DC@OUT} - V_{REF,lo} + V_S) \\ q_{CM_{n,1}} &= C_{CM}(V_{REF,hi} - V_{REF,lo}) \\ q_{INT_{n,1}} &= C_{INT} V_{OUT_{n-1,2}} \end{cases} \tag{A.45}$$

During ϕ_2 the capacitor charges are:

$$\begin{cases} q_{S_{n,2}} &= C_S(V_{REF,hi} - V_{REF,lo}) \\ q_{CM_{n,2}} &= 0 \\ q_{INT_{n,2}} &= C_{INT} V_{OUT_{n,2}} \end{cases} \tag{A.46}$$

Also given is:

$$V_{DC@OUT} = \frac{V_{REF,lo} + V_{REF,hi}}{2} \tag{A.47}$$

$$C_{CM} = Dsty \tfrac{C_S}{2} \tag{A.48}$$

The charge conservation equations are:

$$\begin{cases} q_{S_{n,1}} + q_{CM_{n,1}} + q_{INT_{n,1}} = q_{S_{n,2}} + q_{CM_{n,2}} + q_{INT_{n,2}} \\ q_{INT_{n,1}} = q_{INT_{n-1,2}} \end{cases} \tag{A.49}$$

This readily gives

$$V_{OUT_{n,2}} = V_{OUT_{n-1,2}} + \frac{C_S}{C_{INT}} V_S \tag{A.50}$$

This means that the integrator transfer function is correct.

Appendix B
Calculations for Chapter 4

B.1 Noise Analysis of Low Voltage Current Mirror.

In this section a low frequency noise analysis of the low voltage current mirror is done.The noise source of M_I is modeled as a drain current $i_n = \sqrt{\overline{di_n^2}}$. Referring to Fig. B.1 the nodal equations are:

$$\begin{cases} i_n + (v_2 + v_I)g_{oI} + g_{mI}v_I - \frac{-v_I}{R_S} - i_{fb} = 0 \\ i_n + (v_2 + v_I)g_{oI} + g_{mI}v_I + v_2 g_{oB} = 0 \\ i_{fb} = g_{m2}v_2 \end{cases} \tag{B.1}$$

These yield

$$\begin{cases} v_I \approx \dfrac{i_n}{g_{mI} + g_S} & \text{with} \quad g_S = R_S^{-1} \\ v_2 \approx \dfrac{-i_n}{g_{m2}g_{mI}R_S} & \\ i_{no} \approx \dfrac{-i_n}{g_{mI}R_S} & \end{cases} \tag{B.2}$$

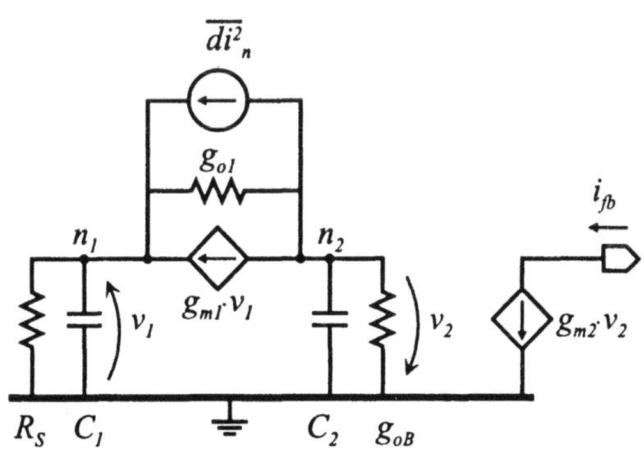

Figure B.1. *The small signal equivalent circuit with noise sources of the low voltage current mirror of Fig. 4.2(d) in open loop configuration.*

This can be further simplified for two special cases.

$$R_S = 0 \quad \Longrightarrow \quad i_o \approx \frac{-i_n}{2} \tag{B.3}$$

$$g_{mI} R_S >> 2 \quad \Longrightarrow \quad i_{no} \approx \frac{-i_n}{g_{mI} R_S} \approx 0 \tag{B.4}$$

The transfer to the output for the noise currents of M_2 and the top current source transistor is not derived analytically. Because the current bias is actually also an input to the current mirror and the noise current from M_2 is plainly injected into the input node, it is clear that they appear unattenuated at the output.

B.2 Noise Analysis of New Input Stage.

In this section a low frequency noise analysis of the new input stage is done. First the noise source of M_{Ia} is treated. The small signal model with noise sources is shown in Fig. B.2 The nodal equations are:

$$\begin{cases} -(2g_{mI} + g_{o2})v_I + i_{nIa} - g_{m2}v_2 = 0 \\ v_2 = r_{oB} g_{mI} v_I - i_{nIa} r_{oB} \end{cases} \tag{B.5}$$

These yield

$$\begin{cases} v_I = \dfrac{1 + \frac{g_{m2}}{g_{oB}}}{g_{mI}\left(2 + \frac{g_{m2}}{g_{oB}}\right) + g_{o2}} i_{nIa} \\ v_2 = \dfrac{1}{g_{oB}} \left(\dfrac{-g_{mI} + g_{o2}}{g_{mI}\left(2 + \dfrac{g_{m2}}{g_{oB}}\right) + g_{o2}} \right) i_n \approx \dfrac{-i_{nIa}}{g_{m2}} \end{cases} \tag{B.6}$$

And consequently

$$i_{no} \approx i_{nIa} \tag{B.7}$$

Also the transfer to i_{Ib} should be investigated, since that current is also directly mirrored to the output in the class AB OTA.

$$i_{Ib} = -g_{mI} v_I = \frac{1 + \dfrac{g_{m2}}{g_{oB}}}{g_{mI}\left(2 + \dfrac{g_{m2}}{g_{oB}}\right) + g_{o2}} i_{nIa} \approx -i_{nIa} \tag{B.8}$$

There are a few other cases to consider. Very straightforward is M_{Ib}. The drain source noise current is injected into the input node of the current buffer structure of M_2, M_{Ia} and M_3. So it appears unattenuated at the output. It can be concluded that M_{Ia} and M_{Ib} have equal noise contributions. The noise current i_{nIa} doesn't really appear at the output (it flows in a loop through the output conductance of M_{Ia}), but it generates a noise current in M_{Ib}. Note that it is necessary to keep the signs in the derived noise currents, because they are required to determine the exact differential transfer.

The noise generated by the top current source i_{nB} (in parallel with Z_{n2}) is also worthwhile deriving analytically.

$$\begin{cases} -(2g_{mI} + g_{o2})v_I - g_{m2}v_2 = 0 \\ v_2 = r_{oB} g_{mI} v_I - i_{nB} r_{oB} \end{cases} \tag{B.9}$$

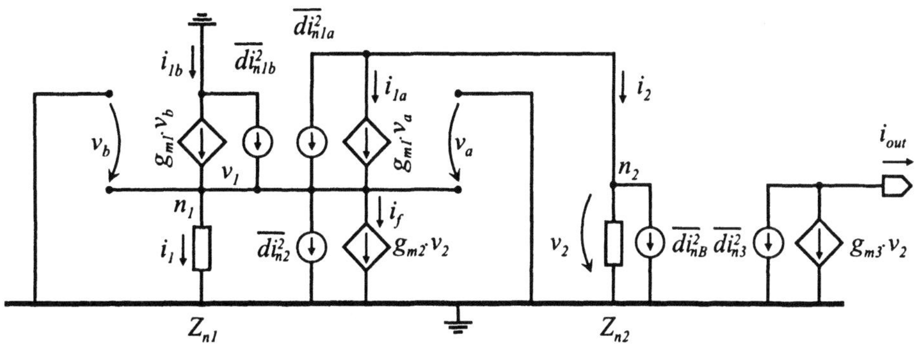

Figure B.2. *The small signal equivalent circuit with noise sources of the new input stage Fig. 4.5.*

These yield

$$\begin{cases} v_1 = \dfrac{\dfrac{g_{m2}}{g_{oB}}}{g_{m1}\left(2+\dfrac{g_{m2}}{g_{oB}}\right)+g_{o2}} i_{nB} \approx \dfrac{i_{nB}}{g_{m1}} \\ v_2 = -\dfrac{1}{g_{oB}}\left(\dfrac{2g_{m1}+g_{o2}}{g_{m1}\left(2+\dfrac{g_{m2}}{g_{oB}}\right)+g_{o2}}\right) i_{nB} \approx \dfrac{-2i_{nB}}{g_{m2}} \end{cases} \tag{B.10}$$

And consequently

$$i_{no} \approx 2i_{nB} \tag{B.11}$$

Also the transfer to i_{1b} should be investigated, since that current is also directly mirrored to the output in the class AB OTA.

$$i_{1b} = -g_{m1}v_1 \approx -i_{nB} \tag{B.12}$$

This is remarkable. The noise current i_{nB} doesn't only transfer to the output by a factor of 2 (4 in power), but it causes a noise current in M_{1b}. Hence, if that current is reused as in the low voltage class AB OTA, it appears at the other output and adds up to the differential output noise current in a correlated manner. So differentially, i_{nB} transfers to the output by a factor of 3 (or 9 in power).

Finally it is straightforward to see that i_{n2} is simply mirrored to the output.

B.3 Derivation of Trifferential Stage Branch Currents

Here the branch currents for the trifferential input stage of Fig. 4.15 are calculated. The small signal equivalent model is given in Fig. B.3. The nodal equation in node n_S is:

$$(g_{m1a}+sC_{GS1a})(v_{in}-v_S)+(2g_{m1b}+s2C_{GS1b})(-v_S)-sC_pv_S \tag{B.13}$$

Note that transistors M_{1b} and M_{1c} are considered equal. Capacitor C_p is the parasitic capacitors at n_S: $C_p = C_{SB1a,b,c} + C_{DB4} + C_{GD4}$. This yields

$$v_S = \frac{(g_{m1a}+sC_{GS1a})}{(g_{m1a}+2g_{m1b})+s(C_{GS1a}+2C_{GS1b}+C_p)} \tag{B.14}$$

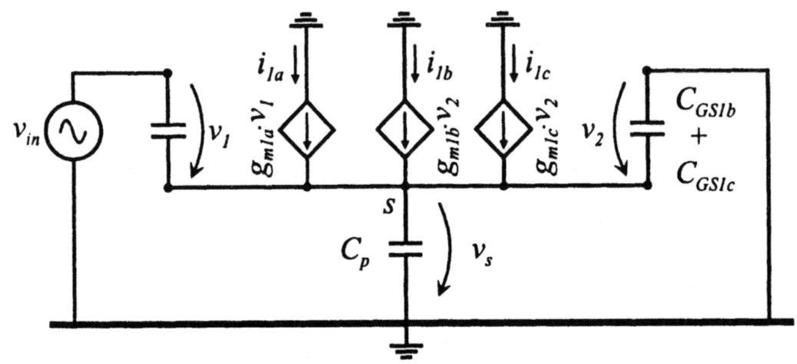

Figure B.3. *The small signal equivalent circuit of the trifferential input stage of Fig. 4.15.*

Then the branch currents can be calculated

$$i_{1a} = g_{m1a}(v_{in} - v_S) \tag{B.15}$$

$$= \left(\frac{g_{m1a}2g_{m1b}}{g_{m1a}+2g_{m1b}}\right)\left(\frac{1+s\left(\frac{2C_{GS1b}+C_p}{2g_{m1b}}\right)}{1+s\left(\frac{C_{GS1a}+2C_{GS1b}+C_p}{g_{m1a}+2g_{m1b}}\right)}\right) \tag{B.16}$$

$$i_{1b} = i_{1c} = g_{m1b}(-v_S) \tag{B.17}$$

$$= -\left(\frac{g_{m1a}g_{m1b}}{g_{m1a}+2g_{m1b}}\right)\left(\frac{1+s\left(\frac{C_{GS1a}}{g_{m1a}}\right)}{1+s\left(\frac{C_{GS1a}+2C_{GS1b}+C_p}{g_{m1a}+2g_{m1b}}\right)}\right) \tag{B.18}$$

Appendix C
Settling Analysis

The purpose of this section is to verify quantitatively the effect on settling behavior of several pole and zero combinations in the transfer function of an opamp in DC feedback configuration. Several cases are examined: a pole-zero pair at a frequency lower than the GBW and assuming real closed loop poles, a pole-zero pair at a frequency higher than the GBW assuming real closed loop poles and assuming complex closed loop poles.

C.1 Pole-Zero Pair with Frequency Lower than Gain-Bandwidth

In this section the effect of pole-zero cancellation is studied for the pole-zero frequency occurring below the GBW frequency. Further, it is also assumed that the closed loop poles are real.

The open loop transfer function of the opamp is then given by

$$H(s) = A\frac{(1+s\tau_z)}{(1+s\tau_{pl})(1+s\tau_{psz2})} \tag{C.1}$$

The closed loop transfer function is

$$\begin{aligned} G(s) &= \frac{H(s)}{1+\mathcal{F}_{dc}H(s)} \\ &= \frac{A}{\mathcal{F}_{dc}A+1}\frac{(1+s\tau_z)}{(1+s(\frac{\tau_{pl}+\tau_{p2}}{\mathcal{F}_{dc}A+1}+\frac{\mathcal{F}_{dc}A}{\mathcal{F}_{dc}A+1}\tau_z)+s^2\frac{\tau_{pl}\tau_{p2}}{\mathcal{F}_{dc}A+1})} \\ &\approx \frac{1}{\mathcal{F}_{dc}}\frac{(1+s\tau_z)}{(1+s\tau_z+s^2\tau_u\tau_{p2})} \end{aligned} \tag{C.2}$$

with

$$\tau_u = \frac{\tau_{pl}}{\mathcal{F}_{dc}A+1} \tag{C.3}$$

The assumption made here is that $\tau_z \gg \dfrac{\tau_{pl}+\tau_{p2}}{\mathcal{F}_{dc}A+1}$. Further the poles are assumed real. This occurs for the condition $D > 0$, which is fulfilled for $\tau_z > 4\tau_u$. Hereby it is also assumed that $\tau_z \approx \tau_{p2}$. The two closed loop poles of the approximated transfer function can be calculated as

$$\omega_{p\alpha,\beta} = \frac{-\tau_z \pm \sqrt{\tau_z^2 - 4\tau_u\tau_{p2}}}{2\tau_u\tau_{p2}} \tag{C.4}$$

The two poles approximate to

$$\omega_{p\alpha} = \frac{(1+\epsilon)}{\tau_u}$$
$$\omega_{p\beta} = \frac{1}{\tau_z}\left(1 - \frac{\tau_u}{\tau_z}\frac{1}{(1+\epsilon)}\right) = \frac{\rho}{\tau_z} \tag{C.5}$$

where $\frac{\tau_z}{\tau_{p2}} = (1+\epsilon)$ and $\rho = \left(1 - \frac{\tau_u}{\tau_z}\frac{1}{(1+\epsilon)}\right)$ So the approximate closed loop transfer function is given by

$$G(s) \approx \frac{1}{\mathcal{F}_{dc}} \frac{(1+s\tau_z)}{(1+s\frac{\tau_u}{(1+\epsilon)})(1+s\frac{\tau_z}{\rho})} \tag{C.6}$$

The step response in the time domain $u(t)$ is obtained by taking the inverse Laplace transform of the transfer function for a step input $1/s$.

$$\begin{aligned} u(t) &= \mathcal{L}^{-1}\left\{G(s)\tfrac{1}{s}\right\} \\ &= \frac{1}{\mathcal{F}_{dc}}\left(1 - \underbrace{\frac{\rho(\tau_u - \tau_z(1+\epsilon))}{\rho\tau_u - (\tau_z(1+\epsilon))}}_{\approx 1 - \frac{\tau_u}{\tau_z}\frac{1}{(1+\epsilon)}} e^{-\frac{(1+\epsilon)}{\tau_u}t} - \underbrace{\frac{\tau_z((\rho-1)(1+\epsilon))}{\rho\tau_u - (\tau_z(1+\epsilon))}}_{\approx -\frac{\tau_u}{\tau_z}\frac{1}{(1+\epsilon)}} e^{-\frac{\rho}{\tau_z}t}\right) \end{aligned} \tag{C.7}$$

The two poles determine two exponential terms in $u(t)$. One term settles fast with a time constant of the order of τ_u and has a large coefficient. The other term has a much lower coefficient, but settles with a time constant of the order of τ_z. This is a slow component since it is assumed in this analysis that the pole-zero pair occurs at a frequency (much) below the GBW. This slow settling term dominates the settling behavior. It is present, because it is not possible to cancel out completely the closed loop pole with the closed loop zero.

It can be concluded that compensating poles that occur below the GBW with zeros causes the settling process to be limited by the time-constant related to the frequency where the cancellation occurs.

C.2 Pole-Zero Pair with Frequency higher than Gain-Bandwidth

In this section the effect of pole-zero cancellation is studied for the pole-zero frequency occurring higher than the GBW frequency. Since for cancellation the zero must have the same sign as the pole it tries to cancel and the pole must be negative for stability, only a negative zero is considered. Further, it is also assumed that the closed loop poles are real.

Equations (C.1) to (C.2) are identical for this case. Now, with the assumption that $\tau_u = \frac{\tau_{pl}}{\mathcal{F}_{dc}A} \ll \{\tau_z, \tau_{p2}\}$ the closed loop transfer function reduces to

$$G(s) = \frac{1}{\mathcal{F}_{dc}} \frac{(1+s\tau_z)}{(1+s\tau_u)(1+s\tau_{p2})} \tag{C.8}$$

Assuming that $\tau_z = \tau_{p2}(1+\epsilon)$ the step response is given by

$$u(t) = \frac{1}{\mathcal{F}_{dc}}\left(1 - \frac{-\tau_{p2}+\tau_z}{-\tau_{p2}+\tau_u} e^{-\frac{t}{\tau_{p2}}} - \frac{-\tau_z+\tau_u}{-\tau_{p2}+\tau_u} e^{-\frac{t}{\tau_u}}\right) \tag{C.9}$$

For a mismatch in the pole-zero pair equal to ϵ, or $\tau_z = \tau_{p2}(1+\epsilon)$, the step response becomes:

$$u(t) = \frac{1}{\mathcal{F}_{dc}}\Big(1 - \underbrace{\frac{\tau_{p2}\,\epsilon}{-\tau_{p2}+\tau_u}}_{\approx \epsilon} e^{-\frac{t}{\tau_{p2}}} - \underbrace{\frac{-\tau_{p2}(1+\epsilon)+\tau_z}{-\tau_{p2}+\tau_u}}_{\approx -(1+\epsilon) \approx -1} e^{-\frac{t}{\tau_u}}\Big) \tag{C.10}$$

There are two settling terms. One term is determined by time-constant corresponding to the *GBW* of the loop. The other by the non-dominant pole and this term settles faster. It also has a coefficient that is proportional to the matching error between the pole and the zero and to the ratio of the *GBW* to the pole-zero frequency. Besides settling faster than the dominant term, this term also has a much smaller coefficient. As the frequency of the pole increases, the effect becomes even more negligible. It can be concluded that pole-zero pairs occurring at a frequency higher than the *GBW* do not degrade settling behavior significantly.

C.3 Pole-Zero Doublet at frequency higher than Gain-Bandwidth

In this section the effect of pole-zero doublets on settling behavior is studied for the pole-zero frequency occurring higher than the *GBW* frequency. Further, it is also assumed that the closed loop poles are real.

Expression (C.9) also holds for any pole and zero combination that occurs at higher frequency than the *GBW* of the loop. So it also holds for a pole-zero doublet, and in that case reduces to:

$$u(t) = \frac{1}{\mathcal{F}_{dc}}\Big(1 - \frac{1}{2}e^{-\frac{t}{\tau_{p2}}} - \frac{1}{2}e^{-\frac{t}{\tau_u}}\Big) \tag{C.11}$$

The dominant settling component is still the one determined by the *GBW*: it is slowest and has . The pole determines a settling component that has a coefficient equal to that of the dominant component but is much faster.

It can be concluded that pole-zero doublets occurring at a frequency higher than the *GBW* do not degrade settling behavior significantly.

C.4 Two Complex Poles and a Zero

In this section the effect on settling behavior of a closed loop transfer function with two complex poles and a zero is studied.

Consider the open loop transfer function of the opamp to be given by

$$H(s) = A\frac{(1+\frac{s}{\omega_z})}{(1+\frac{s}{\omega_{p1}})(1+\frac{s}{\omega_{p2}})} \tag{C.12}$$

The closed loop transfer function is

$$\begin{aligned}
G(s) &= \frac{H(s)}{1+\mathcal{F}_{dc}H(s)} \\
&= \frac{\frac{A\omega_{pl}\omega_{p2}}{\omega_z}(s+\omega_z)}{(s^2+s(\omega_{pl}+\omega_{p2}+\mathcal{F}_{dc}A\frac{\omega_{pl}\omega_{p2}}{\omega_z})+(\omega_{pl}\omega_{p2}+\mathcal{F}_{dc}A\omega_{pl}\omega_{p2}))} \\
&= \frac{\omega_n^2}{\mathcal{F}_{dc}}\frac{1+\frac{s}{\omega_z}}{(s^2+2\zeta\omega_n s+\omega_n^2)}
\end{aligned}$$

with

$$\begin{aligned}
\omega_n^2 &= \omega_{GBW}\omega_{p2} \\
\omega_{GBW} &= A\mathcal{F}_{dc}\omega_{pl} \\
\zeta &\approx \frac{1}{2}(1+\frac{\omega_{GBW}}{\omega_z})\sqrt{\frac{\omega_{p2}}{\omega_{GBW}}}
\end{aligned} \tag{C.13}$$

The step response is then

$$r_u(t) = \left(1+\frac{\sqrt{1-2\zeta\frac{\omega_n}{\omega_z}+\left(\frac{\omega_n}{\omega_z}\right)^2}}{\sqrt{1-\zeta^2}}e^{-\zeta\omega_n t}\sin(\omega_n\sqrt{1-\zeta^2}\,t-\phi)\right)u(t) \tag{C.14}$$

The envelope of the settling error, which is a maximum bound, is then given by

$$\epsilon_S(t) = \frac{\sqrt{1-2\zeta\frac{\omega_n}{\omega_z}+\left(\frac{\omega_n}{\omega_z}^2\right)}}{\sqrt{1-\zeta^2}}e^{-\zeta\omega_n t} \tag{C.15}$$

for $\zeta < 1$, which is the condition for complex poles.

For comparison the envelope of the settling error for a system with no zero is given by

$$\epsilon_S(t) = \frac{1}{\sqrt{(1-\zeta^2)}}e^{-\zeta\omega_n t} \tag{C.16}$$

The presence of a zero, positive or negative, does not change the natural frequency at all. It does have an influence the damping factor ζ, though. It can be verified that a negative zero tends to increase the damping factor, and thus cause real poles. That case is covered by the previous section. As long as ζ stays below unity, a higher value causes better settling. When the damping factor goes above one, though, settling is less fast. A positive zero, on the other hand, causes a drastic decrease of damping factor if its frequency is not much higher than the *GBW*. A lower damping factor means a worse settling.

It can thus be concluded that for complex poles, a negative zero (in the order of the natural frequency) can in general be applied to enhance the settling speed. A positive zero on the other hand must only occur at a frequency much higher than the *GBW* in order not to degrade settling time.

C.5 General Conclusions

This qualitative study of the effect of poles and zeros allows to draw the following conclusions. In the case of pole-zero cancellation at a frequency below the *GBW*, the settling is limited by the time-constant corresponding to the pole-zero frequency. Pole-zero cancellation and pole-zero doublets at a frequency higher than the *GBW* do not significantly degrade settling performance. A zero that occurs in a system with complex poles can enhance settling if it is a negative zero. If the zero is positive, it must be at a high frequency in order not to degrade settling. Overall it can be said that the occurence of many poles and zeros in whatever combination in an amplifier must not lead to the misconception that complete settling is impossible, especially not if they occur at frequencies higher than the *GBW*.

References

[Abo 96] A. Abo, "Low-voltage high-speed high-precision switched-capacitor circuits", Tech. Rep., University of California, Berkeley, Mar. 1996.

[Ada 90] T. Adachi, A. Ishikawa, A. Barlow, and K. Takasuka, "A 1.4V switched capacitor filter", in *Proceedings Custom Integrated Circuits Conference*, 1990, pp. 8.2.1–8.2.4.

[Ard 87] S. H. Ardalan and J. J. Paulos, "An analysis of nonlinear behavior in delta-sigma modulators", *IEEE Transactions on Circuits and Systems*, vol. CAS-34, pp. 593–603, June 1987.

[Au 97] S. Au and B. H. Leung, "A 1.95-V, 0.34-mW, 12-b sigma-delta modulator stabilized by local feedback loops", *IEEE Journal of Solid-State Circuits*, vol. 32, no. 3, pp. 321–328, Mar. 1997.

[Bab 88] J. N. Babanezhad and R. Gregorian, "A rail-to-rail CMOS opamp", *IEEE Journal of Solid-State Circuits*, vol. 23, pp. 1414–1417, Dec. 1988.

[Bas 94] A. Baschirotto, R. Castello, and F. Montecchi, "Design strategy for low-voltage SC circuits", *Electronics Letters*, vol. 30, no. 5, pp. 378–379, Mar. 1994.

[Bas 97a] A. Baschirotto and R. Castello, "A 1V 1.8MHz CMOS Switched-Opamp SC filter with rail-to-rail output swing", in *Proceedings International Solid-State Circuits Conference*, Feb. 1997, pp. 58–56.

[Bas 97b] A. Baschirotto and R. Castello, "A 1V 1.8MHz CMOS Switched-Opamp SC filter with rail-to-rail output swing", *IEEE Journal of Solid-State Circuits*, vol. 32, no. 12, pp. 1979–1986, Dec. 1997.

[Bas 97c] A. Baschirotto and R. Castello, "A 1V CMOS fully-differential Switched-Opamp bandpass $\Sigma\Delta$ modulator", in *Proceedings of the 23rd European Solid-State Circuits Conference*, Sept. 1997, pp. 152–155.

[Bas 98] A. Baschirotto and R. Castello, "A 1V switched-capacitor filters", in *Proceedings Workshop on Advances in Analog Circuit Design*, Apr. 1998.

[Baz 95a] S. S. Bazarjani, T. MacElwee, and M. Snelgrove, "Optimizing the natural mosfets in a 0.5 μm dual poly gate CMOS process for 1V mixed-signal applications", in *Proceedings 25th European Solid State Device Research Conference*, The Hague,The Netherlands, Sept. 1995, pp. 675–678.

[Baz 95b] S. S. Bazarjani and M. Snelgrove, "Low voltage SC circuit design with low-V_T MOSFETs", in *Proceedings IEEE International Symposium on Circuits and Systems*, May 1995, pp. 1021–1024.

[Baz 95c] S. S. Bazarjani, M. Snelgrove, and T. MacElwee, "A 1V switched-capacitor ΣΔ modulator", in *Proceedings IEEE Symposium on Low Power Electronics*, San jose,CA, Oct. 1995, pp. 70–71.

[Bos 88] B. E. Boser and B. A. Wooley, "The design of sigma-delta modulation analog-to-digital converters", *IEEE Journal of Solid-State Circuits*, vol. SC-23, pp. 1298–1308, Dec. 1988.

[Bra 91] B. Brandt and B. Wooley, "A 50-MHz multi-bit sigma-delta modulator for 12-b 2-MHz A/D conversion", *IEEE Journal of Solid-State Circuits*, vol. 26, no. 12, pp. 1746–1756, Dec. 1991.

[Bru 75] L. T. Bruton, "Low-sensitivity digital ladder filters", *IEEE Transactions on Circuits and Systems*, vol. CAS-22, pp. 168–176, Mar. 1975.

[Bur 94] J. B. Burr and J. Shott, "A $200mV$ self-testing encoder/decoder using stanford ultra-low-power CMOS", in *Proceedings International Solid-State Circuits Conference*, feb 1994, pp. 84–85.

[Cal 89] F. Callias, F. H. Salchli, and D. Girard, "A set of four IC's in CMOS technology for a programmable hearing aid", *IEEE Journal of Solid-State Circuits*, vol. 20, no. 2, pp. 301–312, Apr. 1989.

[Can 92] J. C. Candy and G. C. Temes, *Oversampling Methods for A/D and D/A Conversion*, pp. 1–25, IEEE, 1992.

[Car 89] L. R. Carley, "A noise-shaping coder topology for 15+ bit converters", *IEEE Journal of Solid-State Circuits*, vol. 28, no. 2, pp. 267–273, Apr. 1989.

[Cas 85] R. Castello and P. R. Gray, "A high-performance micropower switched-capacitor filter", *IEEE Journal of Solid-State Circuits*, vol. SC-20, no. 6, pp. 1122–1132, Dec. 1985.

[Cas 91] R. Castello and L. Tomasini, "1.5V high-performance SC filters in BiCMOS technology", *IEEE Journal of Solid-State Circuits*, vol. SC-26, no. 7, pp. 930–936, July 1991.

[Cat 89] T. Cataltepe, A. R. Kramer, L. E. Larsen, G. C. Temes, , and R. H. Walden, "Digitally corrected multi-bit ΣΔ data converters", in *Proceedings IEEE International Symposium on Circuits and Systems*, 1989, pp. 647–650.

[Che 90] J. Y. Chen, *CMOS Devices and Technology for VLSI*, Prentice-Hall, Inc., Englewood Cliffs, NJ, 1990.

[Che 95] F. Chen and B. H. Leung, "A high resolution multibit sigma-delta modulator with individual level averaging", *IEEE Journal of Solid-State Circuits*, vol. 30, no. 4, pp. 453–460, Apr. 1995.

[Cho 95] T. B. Cho and P. R. Gray, "A 10 b, 20 Msample/s, 35 mW pipeline A/D converter", *IEEE Journal of Solid-State Circuits*, vol. 30, pp. 166–172, Mar. 1995.

[Coc 32] J. D. Cockcroft and E. T. Walton, "Production of high velocity positive ions", in *Proceeding Royal Society*, 1932, vol. 136 of *A*, pp. 619–630.

[Cra 95] S. Crapanzano and D. A. Johns, "2V fully-differential SC integrator in standard CMOS", *Electronics Letters*, vol. 31, no. 23, pp. 1995–1996, Nov. 1995.

[Cro 94] J. Crols and M. Steyaert, "Switched-Opamp: An approach to realize full CMOS switched-capacitor filters at very low power supply", *IEEE Journal of Solid-State Circuits*, vol. 29, no. 8, pp. 936–942, Aug. 1994.

[Dic 76] J. F. Dickson, "On-chip high-voltage generation in MNOS integrated circuits using an improved voltage multiplier technique", *IEEE Journal of Solid-State Circuits*, vol. SC-11, no. 3, pp. 374–378, June 1976.

[Eyn 93a] F. O. Eynde and W. Sansen, *Analog interfaces for digital signal processing systems*, Kluwer Academic Publishers, 1993.

[Eyn 93b] F. O. Eynde and W. Sansen, *Analog interfaces for digital signal processing systems*, chapter 4.3.g, p. 179, In [Eyn 93a], 1993.

[Fav 97] P. Favrat, P. Deval, and M. J. Declerq, "A new high efficiency CMOS voltage doubler", in *Proceedings Custom Integrated Circuits Conference*, 1997, pp. 12.2.1–12.2.4.

[Fav 98] P. Favrat, P. Deval, and M. J. Declerq, "A new high efficiency CMOS voltage doubler", *IEEE Journal of Solid-State Circuits*, vol. 33, no. 3, pp. 410–416, Mar. 1998.

[Fee 91] O. Feely and L. O. Chua, "The effect of integrator leak in sigma-delta modulation", *IEEE Transactions on Circuits and Systems*, vol. 38, no. 11, pp. 1293–1305, Nov. 1991.

[Fis 82] J. H. Fischer, "Noise sources and calculation techniques for switched capacitor filters", *IEEE Journal of Solid-State Circuits*, vol. SC-17, no. 4, pp. 742–752, Aug. 1982.

[Fis 87] J. A. Fischer and R. Koch, "A highly linear CMOS buffer amplifier", *IEEE Journal of Solid-State Circuits*, vol. SC-22, no. 3, June 1987.

[Fis 97] G. Fisher and A. J. Davis, "Alternative topologies for sigma-delta modulators – a comparative study", *IEEE Transactions on Circuits and Systems II*, vol. 44, no. 10, pp. 789–797, Oct. 1997.

[Gei 82] R. L. Geiger and E. Sanchez-Sinencio, "Operational amplifier gain-bandwidth product effects on the performance of switched-capacitor networks", *IEEE Transactions on Circuits and Systems*, vol. CAS-29, no. 2, pp. 96–106, Feb. 1982.

[Gob 80] C. Gobet and A. Knob, "Noise generated in switched capacitor networks", *Electronics Letters*, vol. 16, no. 19, pp. 734–735, Sept. 1980.

[Gob 83] C.-A. Gobet and A. Knob, "Noise analysis of switched capacitor networks", *IEEE Transactions on Circuits and Systems*, vol. CAS-30, no. 1, pp. 37–43, Jan. 1983.

[Gri 96] J. Grilo, E. MacRobbie, R. Halim, and G. Temes, "A 1.8V 94dB dynamic range sigma-delta modulator for voice applications", in *Proceedings International Solid-State Circuits Conference*, Feb. 1996, pp. 230–231.

[Hai 91] A. Hairapetian, G. D. Temes, and Z. Zhang, "Multibit sigma-delta modulator with reduced sensitivity to DAC nonlinearity", *Electronics Letters*, vol. 27, pp. 990–99, May 1991.

[Har 78] F. Harris, "On the use of windows for harmonic analysis with the dft", *Proceedings of the IEEE*, vol. 66, no. 1, Jan. 1978.

[Kin 97] P. R. Kinget and M. S. J. Steyaert, "A 1-GHz CMOS up-conversion mixer", *IEEE Journal of Solid-State Circuits*, vol. 32, pp. 370–376, Mar. 1997.

[Kru 83] F. Krummenacher, H. Pinier, and A. Guillaume, "Higher sampling rates in SC circuits by on-chip clock-voltage multiplication", in *Proceedings European Solid-State Circuits Conference*, Sept. 1983, pp. 123–126.

[Les 90] T. C. Leslie and B. Singh, "An improved sigma-delta modulator architecture", in *iscas*, 1990, pp. 372–375.

[Leu 92] B. Leung, "Multibit sigma-delta A/D converters incorporating a novel class of dynamic element matching techniques", *IEEE Transactions on Circuits and Systems II*, vol. 39, no. 1, pp. 35–51, Jan. 1992.

[L&S 94] K. R. Laker and W. M. Sansen, *Design of Analog Integrated Circuits and Systems*, McGraw-Hill, New York, 1994.

[L&S a] K. R. Laker and W. M. Sansen, *Design of Analog Integrated Circuits and Systems*, chapter 1.3.1, pp. 14–15, In (I-9) [L&S 94], 1994.

[L&S b] K. R. Laker and W. M. Sansen, *Design of Analog Integrated Circuits and Systems*, chapter 1.2.3 and 1.2.4, pp. 7–10, In (1-4a),(1-6a),(1-6b),(1-6d) and (1-7a) [L&S 94], 1994.

[L&S c] K. R. Laker and W. M. Sansen, *Design of Analog Integrated Circuits and Systems*, chapter 3.3.1, p. 205, In [L&S 94], 1994.

[L&S d] K. R. Laker and W. M. Sansen, *Design of Analog Integrated Circuits and Systems*, chapter 6 - appendix 1, pp. 625–626, In [L&S 94], 1994.

[Mar 81] K. Martin and A. S. Sedra, "Effects of the op amp finite gain and bandwidth on the performance of switched-capacitor filters", *IEEE Transactions on Circuits and Systems*, vol. CAS-28, no. 8, pp. 822–829, Aug. 1981.

[Mar 97] A. Marques, V. Peluso, M. Steyaert, and W. Sansen, "Optimal coefficients for cascade ΔΣ modulators", in *Proceedings IEEE International Symposium on Circuits and Systems*, Hong-Kong, June 1997, pp. 61–64.

[Mar 98a] A. Marques, V. Peluso, M. Steyaert, and W. Sansen, "Analysisof the tradeoffs between bandwidth,resolution and power in ΔΣ a/d converters", in *Proceedings of the 5th International Conference on Electronic, Circuits and Systems*, Lisbon, Sept. 1998.

[Mar 98b] A. Marques, V. Peluso, M. Steyaert, and W. Sansen, "Optimal parameters for $\Delta\Sigma$ modulator topologies", *IEEE Transactions on Circuits and Systems II*, vol. 45, no. 9, pp. 1232–1241, Sept. 1998.

[Mat 87] Y. Matsuya, K. Uchimura, A. Iwata, T. Kobayashi, M. Ishikawa, and T. Yoshitome, "A 16-bit oversampling A-to-D conversion technology using triple-integration noise shaping", *IEEE Journal of Solid-State Circuits*, vol. SC-22, pp. 921–929, Dec. 1987.

[Mat 94] Y. Matsuya and J. Terada, "1V power supply, low-power consumption A/D conversion technique with swing suppression noise shaping", *IEEE Journal of Solid-State Circuits*, vol. 29, no. 12, pp. 1524–1525, Dec. 1994.

[Mat 97] Y. Matsuya and J. Terada, "1.2–V,16–bit audio A/D converter with suppressed latch error noise", 1997.

[Moe 97] N. Moeneclaey and A. Kaiser, "Design techniques for high-resolution current-mode sigma-delta modulators", *IEEE Journal of Solid-State Circuits*, vol. 32, no. 7, pp. 953–958, July 1997.

[Nak 91] Y. Nakagome, H. Tanaka, K. Takeuchi, E. Kume, Y. Watanabe, T. Kaga, Y. Kawamoto, F. Murai, R. Izawa, D. Hisamoto, T. Kisu, T. Nishida, E. Takeda, and K. Itoh, "An experimental 1.5-V 64-Mb DRAM", *IEEE Journal of Solid-State Circuits*, vol. 26, no. 4, pp. 465–470, Apr. 1991.

[Nej 93] M. Nejad and G. Temes, "A high resolution multibit $\Sigma\Delta$ adc with digital correction and relaxed amplifier requirements", *IEEE Journal of Solid-State Circuits*, vol. 28, no. 6, pp. 648–660, June 1993.

[Nic 94] G. Nicollini, A. Nagari, P. Conflonieri, and C. Crippa, "A -80 dB THD $4V_{pp}$ switched capacitor filter for 1.5V battery operated systems", in *Proceedings European Solid-State Circuits Conference*, 1994, pp. 104–107.

[Pel 97a] V. Peluso, A. Marques, M. Steyaert, and W. Sansen, "Optimal coefficients for single loop $\Delta\Sigma$ modulators", in *Proceedings IEEE International Symposium on Circuits and Systems*, Hong-Kong, June 1997, pp. 57–60.

[Pel 97b] V. Peluso, M. Steyaert, and W. Sansen, "A Switched Opamp $1.5V$-$100\,\mu W$ $\Delta\Sigma$ modulator with 12 bits dynamic range", in *Proceedings of the 22nd European Solid-State Circuits Conference*, Neuchâtel Switzerland, Sept. 1996, pp. 256–259.

[Pel 97c] V. Peluso, M. Steyaert, and W. Sansen, "A 1.5V-100 μW $\Delta\Sigma$ modulator with 12-b dynamic range using the Switched-Opamp technique.", *IEEE Journal of Solid-State Circuits*, vol. 32, no. 7, pp. 943–952, July 1997.

[Pel 97d] V. Peluso, P. Vancorenland, M. Steyaert, and W. Sansen, "900 mV class AB OTA for Switched Opamp applications", *Electronics Letters*, vol. 33, no. 17, pp. 1455–1456, Aug. 1997.

[Pel 98a] V. Peluso, P. Vancorenland, A. Marques, M. Steyaert, and W. Sansen, "A 900 mV low-power $\Delta\Sigma$ a/d converter with 77 dB dynamic range", *IEEE Journal of Solid-State Circuits*, vol. 33, no. 12, Dec. 1998, in press.

[Pel 98b] V. Peluso, P. Vancorenland, A. Marques, M. Steyaert, and W. Sansen, "A 900 mV - 40 μW $\Delta\Sigma$ modulator with 77dB dynamic range", in *Proceedings International Solid-State Circuits Conference*, San Francisco, Feb. 1998, pp. 68–69.

[Pla 79] R. J. van de Plassche and D. Goedhart, "A monolithic 14-bit D/A converter", *IEEE Journal of Solid-State Circuits*, vol. SC-14, no. 3, pp. 552–556, June 1979.

[Rab 96] S. Rabii and B. A. Wooley, "A 1.8V, 5.4mW, digital-audio sigma-delta modulator in 1.2 μm CMOS", in *Proceedings International Solid-State Circuits Conference*, Feb. 1996, pp. 228–229.

[Rab 97] S. Rabii and B. A. Wooley, "A 1.8V digital-audio sigma-delta modulator in 0.8 μm CMOS", *IEEE Journal of Solid-State Circuits*, vol. 32, no. 6, pp. 783–796, June 1997.

[Rib 91] D. B. Ribner, "A comparison of modulator networks for high-order oversampled sigma delta analog-to-digital converters", *IEEE Transactions on Circuits and Systems*, vol. CAS-38, no. 2, pp. 145–159, Feb. 1991.

[Rij 93] J. Rijns, "54Mhz switched-capacitor video channel equaliser", *Electronics Letters*, vol. 29, pp. 2181–2182, Dec. 1993.

[Rob 92] A. Robertini and W. Guggenbühl, "Errors in SC circuits derived from linearly modeled amplifiers and switches", *IEEE Transactions on Circuits and Systems I*, vol. 39, no. 2, pp. 93–101, Feb. 1992.

[Roo 96] R. Rovers, *High Speed A/D Converters in Standard CMOS Technology*, chapter 3.4.4, p. 59, Mar. 1996.

[Sau 96] J. Sauerbrey and M. Mauthe, "12 bit - 1mW sigma-delta modulator for 2.4V supply voltage", in *Proceedings of the 21th European Solid-State Circuits Conference*, Sept. 1995, pp. 138–141.

[Sen 97] D. Senderowitz, G. Nicollini, S. Pernici, A. Nagari, P. Confalonieri, and C. Dallavalle, "Low-voltage double-sampled $\Sigma\Delta$ converters", in *Proceedings International Solid-State Circuits Conference*, Feb. 1997, pp. 210–211.

[SIA 97] "The national technology roadmap for semiconductors", Tech. Rep., Semiconductor Industry Association, 1997.

[Ste 88] M. Steyaert and V. Commino, "High-speed accurate CMOS comparator", *Electronics Letters*, vol. 24, no. 16, pp. 1027–1028, Aug. 1988.

[Ste 93] M. Steyaert, J. Crols, and S. Gogaert, "Switched-Opamp, a technique for realising full CMOS switched-capacitor filters at very low voltages", in *Proceedings of the 19th European Solid-State Circuits Conference*, Sept. 1993, pp. 178–181.

[Sun 92] S. W. Sun, "A fully complementary BiCMOS technology for sub-half-micrometer microprocessor applications", *IEEE Transactions on Electron Devices*, vol. 39, no. 12, pp. 2733–2738, Dec. 1992.

[Tan 95] N. Tan, "1.2-V 0.8-mW si $\Delta\Sigma$ A/D converter in standard digital CMOS process", in *Proceedings of the 21th European Solid-State Circuits Conference*, Sept. 1995, pp. 150–153.

[Vit 90] E. Vittoz, "Future of analog in the VLSI environment", in *Proceedings IEEE International Symposium on Circuits and Systems*, May 1990, pp. 1372–1375.

[Way 92] D. Wayne, M. Rives, T. Huynh, D. Preves, and J. Newton, "A single-chip hearing aid with one volt switched-capacitor filters", in *Proceedings Custom Integrated Circuits Conference*, 1992, pp. 7.5.1–7.5.4.

[Wit 89] J. S. Witters, G. Groeseneken, and H. E. Maes, "Analysis and modeling of on-chip high-voltage generator circuits for use in EEPROM circuits", *IEEE Journal of Solid-State Circuits*, vol. 24, no. 5, pp. 1372–1380, Oct. 1989.

[Yin 94a] G. Yin, *High Performance Analog-to-Digital Converters using Cascaded $\Sigma\Delta$ Modulators*, chapter 4.7, pp. 62–69, January 1994.

[Yin 94b] G. Yin and W. Sansen, "A high-frequency and high-resolution fourth-order sigma-delta A/D converter in BiCMOS technology", *IEEE Journal of Solid-State Circuits*, vol. 29, no. 8, pp. 857–865, Aug. 1994.

[Yua 97] J. Yuan and C. Svensson, "New single-clock CMOS latches and flipflops with improved speed and power savings", *IEEE Journal of Solid-State Circuits*, vol. 32, no. 1, pp. 62–69, Jan. 1997.

[Zwa 96a] E. J. van der Zwan and C. Dijkmans, "A 0.2-mW CMOS $\Sigma\Delta$ modulator for speech coding with 80 dB dynamic range", *IEEE Journal of Solid-State Circuits*, vol. 31, no. 12, pp. 1873–1880, Dec. 1996.

[Zwa 96b] E. J. van der Zwan and E. C. Dijkmans, "A 0.2mW CMOS sigma-delta modulator for speech coding with 80 dB dynamic range", in *Proceedings International Solid-State Circuits Conference*, Feb. 1996, pp. 232–233.

[Zwa 97] E. J. van der Zwan, "A 2.3mW CMOS sigma-delta modulator for audio applications", in *Proceedings International Solid-State Circuits Conference*, Feb. 1997, pp. 220–221.

About the Authors

Vincenzo Peluso was born in Genk, Belgium, in 1967. He received the M.Sc. and Ph.D. degree in electrical engineering in 1992 and 1998 from the Katholieke Universiteit (K.U.) Leuven, Belgium. From 1993 to 1997 he was a research assistant at the ESAT-MICAS laboratory at K.U. Leuven. His dissertation was on the design of low-voltage, low-power design of CMOS $\Delta\Sigma$ A/D Converters. His research interests are low-voltage, low-power analog design techniques and $\Delta\Sigma$ modulators. Since 1998 he has been employed by QUALCOMM, Inc. in San Diego, California. There he is working on the design of analog chipsets for wireless communications applications.

Michel Steyaert was born in Aalst, Belgium, in 1959. He received the M.S.E.E. degree in electrical-mechanical engineering and the Ph.D. degree in electronics from the Katholieke (K.U.) Leuven, Belgium, in 1983 and 1987, respectively.

From 1983 to 1986 he worked as a Research Assistant at the laboratory ESAT at K.U. Leuven, having obtained a Belgian National Foundation for Industrial Research (IWONL) fellowship. In 1987, he was an IWONL Project Researcher, Responsible for several industrial projects in the field of analog micropower circuits at the ESAT Laboratory. In 1988, he was appointed Research Associate by the National Fund of Scientific Research, Belgium, then in 1992 was appointed Senior Research Associate, and then Research Director at the Laboratory ESAT in 1996. Between 1989 and 1996, he was also a part-time Associate Professor, and since 1997, an Associate Professor at K.U. Leuven. His current research interests include high-performance and high-frequency analog integrated circuits for telecommunications systems and analog signal processing.

Dr. Steyaert received the 1990 European Solid-State Circuits Conference Best Paper Award, the 1995 and 1997 ISSCC Evening Session Award, and the 1991 NFWO Alcatel-Bell-Telephone award for innovative work in integrated circuits for telecommunications.

Willy Sansen was born in Poperinge, Belgium, on May 16, 1943. He received the M.Sc. degree in electrical engineering from the Katholieke Universiteit (K.U.) Leuven, Belgium, in 1967 and the Ph.D. degree in electronics from the University of California, Berkeley, in 1972.

In 1968, he was a Research Assistant at K.U. Leuven. In 1971, he was a Teaching Fellow at the University of California, Berkeley. In 1972, he was appointed Research Associate at the ESAT Laboratory of K.U. Leuven by the National Fund of Scientific Research, Belgium, where he has been a full professor since 1981. During 1984-1990, he was the head of the Electrical Engineering Department. In 1978, he spent the winter semester at Stanford University as a Visiting Assistant Professor. He was a Visiting Professor in 1981 at the Federal Technical University of Lausanne, then in 1995 at the University of Pennsylvania, and in 1994 at the Technical University of Ulm. He has been involved in design automation and in numerous analog integrated circuit designs for telecommunications, consumer electronics, medical applications, and sensors. He has been supervisor of 40 Ph.D. dissertations in these fields. He has authored and co-authored more than 400 papers in international journals and conference proceedings, in addition to seven books, including *Design of Analog Integrated Circuits and Systems* (with. K. Laker). He is a member of several editorial committees of journals and program committees of conferences, such as ISSCC and ESSCIRC.

Index